Pelargoniums
of
Southern Africa

Pelargoniums of Southern Africa

Text
J.J.A. van der Walt D. Sc., B. Ed.
Senior Lecturer in Botany,
University of Stellenbosch
Illustrations
Ellaphie Ward-Hilhorst

PURNELL
CAPE TOWN · JOHANNESBURG · LONDON

Published by Purnell & Sons S.A. (Pty) Ltd.
97 Keerom Street Cape Town

ISBN 0 86843 006 4

COLOUR REPRODUCTION : HIRT & CARTER CAPE TOWN
TYPESETTING : SPARHAM AND FORD CAPE TOWN
PRINTED AND BOUND IN SOUTH AFRICA
BY CREDA PRESS (PTY) LTD. CAPE TOWN

Foreword

by H. B. Rycroft

M.Sc., B.Sc. (For.), Ph.D., F.L.S., F.R.S.S.Af., Hon. F.R.H.S.,
Hon. F. Inst. P. & R.A. (S.A.)
Director of the National Botanic Gardens of South Africa and
Harold Pearson Professor of Botany at the University of Cape
Town.

Being personally and professionally acquainted with both the author and the artist of this work on Pelargonium, I am in a unique and privileged position to introduce it to the ever-increasing number of people to whom it will be of interest and value. This I do with the greatest pleasure.

I am most impressed and gratified to see the number of books now being published on the South African flora. This flora has attracted the attention and interest of botanists, horticulturists and plant lovers for several centuries. One group which, up to now, however, has not received special attention includes the pelargoniums or the so-called "geraniums"

Every year garden "geraniums" by the million are being produced by nurserymen for adornment and decoration in homes, offices, streets, public parks and botanic gardens throughout the world. They have been mainly derived or bred from the wild species of Pelargonium indigenous to South Africa. While these cultivars have unquestioned horticultural merit their progenitors are well worth special attention. In fact, many of the natural species have undoubted beauty in their pure simplicity of form and colour and should also be accommodated in the home garden.

The last taxonomic work on the genus was undertaken by R. Knuth a German botanist of Berlin, in 1912. Now, 65 years later, a South African botanist, Dr J.J.A. van der Walt of the University of Stellenbosch, is engaged on a complete revision of Pelargonium. He is well qualified to do this work because of his intimate contact with the plants in the field.

Fifty natural species are described. Each description is commenced with a basic nomenclatural exposition, followed by the diagnostic features of the species. Use of technical terms has been limited to a minimum in dealing with habit, flowers and leaves. Notes on the cultivation, ecology, medicinal uses and relevant historical data are given for each species.

We are indeed fortunate in having an authoritative text written by a botanist who has made a special study of the genus. Equally fortunate are we that Ellaphie Ward-Hilhorst is the artist. Her work is remarkably reminiscent of the masterpieces of the great botanical artists of the nineteenth century. Each species is superbly illustrated with a life size water-colour drawing made from authenticated live material. In many cases these are the first colour illustrations ever to be made of pelargoniums in Southern Africa.

This work, with which I am happily associated, will be of interest not only to botanists but also to nature lovers, horticulturists and artists.

BRIAN RYCROFT
May 1977

Acknowledgements

Contributions to this work have been submitted by Mrs. D.A. Boucher, Mr. P.J. Drijfhout, Miss L. Hugo, Miss E.M. Marais and Miss C.M. Schonken of the Botany Department, University of Stellenbosch. I wish to extend my sincere thanks to all these co-workers on the Geraniaceae project. The financial support by the C.S.I.R. and University of Stellenbosch to this project is also acknowledged with appreciation.

We are grateful to the curators and staff of the South African and overseas herbaria in making available to us the loan of valued herbarium material and literature. The illustrator wishes to acknowledge the generous encouragement and assistance of Professor H.B. Rycroft, Dr. John Rourke and his staff of the Compton Herbarium, as well as the encouragement and patience of her family and friends.

We are indebted to Mrs. J.M. Serdyn for typing the manuscript and to Mrs. A.E. Cillié for scrutinizing the text.

J.J.A. van der Walt.

Preface

Geraniums, as they are popularly known, must surely be among the most cosmopolitan, well-loved garden and pot plants. They are artificial hybrids developed by plant breeding methods from plants botanically known as the genus *Pelargonium* of the family Geraniaceae. *Pelargonium* forms a typical element of the Cape Flora and a large majority of species is confined to the winter rainfall area of the Cape Province, Republic of South Africa.

The vernacular use of the name geranium for a plant which should correctly be called a pelargonium, has led to general confusion, as the genus *Pelargonium* differs markedly from the genus *Geranium*.

As confusing is the misnomer "malva" in the Afrikaans language, which gives the impression that the plant is a member of the Malvaceae and not the Geraniaceae. The origin of the name "malva" is uncertain. The word "malve" was used in the Journal of Commander Jan van Riebeeck, dated 24 November 1659. Reference to "malva" was also made in the "Hortus medici Amstelaedamensis Plantae rariores et exoticae" of Casper Commelin (1706). The Latin quotation as translated by Mia Karsten (1951) reads: "This Geranium sprang up from seeds, sent from Africa in 1700 by the most Honourable Governor of the Cape of Good Hope, Wilhelm Adriaan van der Stel, under the name Geranium with the leaf of Malva from the Attaqua region" (i.e. south of the present Oudtshoorn). It may be concluded that the word "malva" gradually infiltrated into the common vocabulary of the laymen.

Pelargoniums of the Cape attracted the attention of travellers since the years of exploration around the coast of the African continent. Plants were brought to Europe and England in the early years of the 17th century to be planted in estate gardens. *Pelargonium triste* was one of the first species to be transplanted to England. We learn of the "Sweet Indian Stork's Bill" or in Latin of the "Geranium Indicum odoratum" from the revised edition of Gerard's Herbal by Thomas Johnson (1633). The name given to this species suggests that the plant had been brought from India.

The Dutch East India Chartered Company which was founded in 1602 sent their trading vessels around the Cape of Storms to the Spice Islands of the Far East. These ships called at the Cape for supplies of fresh drinking water, so necessary on their long journey. After the Dutch had decided to establish a naval station and garrison post at Table Bay under Commander Jan van Riebeeck in 1652, the settlement rapidly became an important stopping post for refreshments between the Netherlands and India. Consequently more travellers stayed over en route to India. Such a visitor was Paul Hermann, a Dutch botanist, who achieved fame as professor of Botany at the University of Leyden in August 1680. He collected *Pelargonium cucullatum* on the slopes of Table Mountain in the year 1672. By 1690 this species was established in the Royal Botanical Gardens at Kew, England, while the gardens of Leyden had ten species of pelargonium on record by 1687.

A pupil of Paul Hermann, Heinrich Bernhard Oldenland, was appointed as gardener at the Cape because of his knowledge of plants and herbs. His compiled list of the indigenous plants in his garden, included 22 pelargoniums, and he also started his own herbarium with Latin descriptions of the plants. After his death in 1795, this excellent collection passed into the herbarium of the Burmans, while even earlier in 1711 James Petiver acquired dried specimens for the Sloane Herbarium in England. From the correspondence of W.A. van der Stel, governor at the Cape from 1699 to 1706, it transpires that he forwarded a parcel of plants to Holland. He introduced *Pelargonium peltatum* to Dutch horticulture in 1700.

As the pelargoniums obtained from South Africa became known in floriculture, keen botanists and artists were inspired to describe and make drawings of them. In 1732 Johan Jakob Dillenius, a German-born English botanist who was the first professor of botany at Oxford, published his work "Hortus Elthamensis". It covers the plants grown in the garden of Sharard at Eltham in the county of Kent, England. He described and made excellent black and white illustrations of *P. inquinans, P. vitifolium, P. carnosum, P. papilionaceum, P. cucullatum, P. fulgidum* and *P. odoratissimum* for this unique work.

It serves to note that at this stage binomial nomenclature for the naming of plants was not yet in practice, and at that time all the pelargoniums were placed in the genus *Geranium.*

Another work of merit called the "Rariorum Africanorum Plantarum" by the Dutch botanist, Johannes Burman, was published in 1738. The drawings for the engravings in black and white in this publication, were made by Hendrik Cladius during his visit to the Cape. It is interesting to note that Burman used the name *Pelargonium* for the species. Linnaeus and successive botanists persisted in naming all the pelargoniums as *Geranium* species until the French botanist, Charles-Louis L'Heritier de Brutelle designated them as belonging to the genus *Pelargonium* in 1789.

A brisk trade in the export of pelargoniums from South Africa to England ensued during the closing years of the 18th century and the first two decades of the 19th century. The trade stimulus came partly from the newly developed methods of cultivating plants in hothouses in England, and partly through the political situation following the British occupation of the Cape.

It was already well known that pelargoniums could easily be cultivated from seed or cuttings. Plant breeding methods yielded quick and good results, so that large numbers of hybrids were developed. Some of these hybrids were unattractive and by selection were discarded in course of time. Clifford (1970) is of opinion that the diagnostic features of twenty natural species of *Pelargonium* at the most, can be traced in our modern hybrids. The tremendous popularity of pelargoniums and geraniums as pot or garden plants led to the establishment of many Pelargonium and Geranium societies in many different countries of the world. One may mention here the British Pelargonium and Geranium Society, the Australian Geranium Society, the South African Pelargonium and Geranium Society and the International Geranium Society. Today cultivars are grown on an unprecedented scale. New experimental methods are tried out to cultivate new varieties. The following paragraph is quoted from Mastalerz's (1971) publication: "Geraniums represent one of the most exciting commercial flower crops in the U.S.A. No other flower has shown a greater rate of increase in dollar value to commercial floriculture and better performance for purchasers during the last twenty years". In both the publications of Clifford (1970) and Mastalerz (1971) much attention is given to the aspect of growing pelargoniums for both their decorative and commercial value.

Certain species of *Pelargonium,* viz. *P. radens, P. graveolens, P. capitatum* and *P. odoratissimum,* yield geranium oil which contains geraniol and citronellol, substitutes for the expensive attar of roses in the perfume trade. Pelargoniums are cultivated for their oil yield in the south of Europe, on the island of Reunion and in several African states.

The medicinal value of pelargoniums, supposed to give relief in cases of diarrhoea and dysentery, was known to the earliest tribes of Southern Africa.

The family Geraniaceae, which includes *Pelargonium,* formerly circumscribed a much wider range of plants. Hutchinson (1969) and other botanists hold that the Geraniaceae should only include the five genera *Geranium, Erodium, Monsonia, Sarcocaulon* and *Pelargonium.* The common diagnostic feature of all the members of the five genera is the typical elongated fruit or schizocarp which has five single locules (mericarps), each with one seed. The popular name of "Stork's Bill" or "Crane's Bill" is derived from the shape of the schizocarp, which resembles a stork's bill. Representatives of all five genera are found in South Africa.

Geranium: Geranos (Greek), a crane; "crane's bill" refers to the rostrum of the schizocarp; resembling the bill of a crane. This genus includes about 400 species occurring mainly in temperate regions. According to Dyer (1974) about 18 species are found in South Africa. The flower is actinomorphic or regular, with ten fertile (i.e. pollen producing) stamens.

Erodium: Erodios (Greek), a heron; Heron's bill refers to the rostrum of the fruit resembling the bill of a heron. This genus includes about 60 species, most of which are distributed throughout the Mediterranean region. The larger number of species found in South Africa are apparently exotic. The flowers of *Erodium* are actinomorphic with only five fertile stamens.

Monsonia was named in honour of Lady Anne Monson, a great grand-daughter of Charles II, and a prominent figure in Calcutta Society. This genus includes 40 species, widely distributed throughout the African continent, with a few species in India as well. About 30 species have a South African distribution according to Dyer (1974). The flowers are actinomorphic and have 15 fertile stamens. The plants are herbaceous.

Sarcocaulon: Sarco (Greek), fleshy; caulon (Greek) stem; Bushman candles, "Boesmankerse". This genus is restricted to Southern Africa, occurring in the drier regions of the Cape Province, of South West Africa and Angola. It includes about 15 species. The flowers are actinomorphic with 15 fertile stamens. The plants have fleshy stems, often with persisting spiny petioles.

Pelargonium: Pelargos (Greek), stork; Stork's Bill refers to the rostrum of the schizocarp which resembles the bill of a stork. The genus includes more than 200 natural species and numerous artificial hybrids. The larger majority of the natural species is South African, while a few species occur in Tropical Africa, Syria, Australia and on a few islands in the Indian Ocean. Of importance is the fact that the flowers of the genus are zygomorphic, with a nectar spur, and with not more than seven stamens out of the ten being fertile. It is very easy to identify the South African genera of the Geraniaceae by means of the following key:

1 Flowers actinomorphic and without a nectar spur
  2 Stamens 10
    3 Fertile stamens 10 ....*Geranium*
    3 Fertile stamens 5 ....*Erodium*
  2 Stamens 15
    4 Stems herbaceous without spines ....*Monsonia*
    4 Stems succulent, generally with spines ....*Sarcocaulon*
1 Flowers zygomorphic with a nectar spur *Pelargonium*

Contents

The plate illustrating each species in life size,
appears opposite the relevant text

The taxonomic and nomenclatural history of Pelargonium

A wealth of interesting and closely interrelated facts emerges from the study of the taxonomic history of the genus *Pelargonium.*

Against a background of political unrest in Europe, of wars and of governments displaced in England, France, Holland and the Cape of Good Hope, the study stretches over three centuries up to the present day. It discloses the hardships endured by the early explorers and pioneers of Southern Africa, and the privations suffered by the botanists on their inland travels.

Many of the early botanists, such as Linnaeus and the Burmans, who published on this genus, never visited South Africa and thus never had the opportunity to study the plants in their natural environment. The only material available to these botanists, was a limited number of herbarium specimens and cultivated plants from the gardens of Europe and England.

The nomenclatural history of the genus *Pelargonium* thus proves to be complicated. A brief resumé of the more important botanists who were concerned with the taxonomy and nomenclature of the genus, will be given:

Linnaeus, Carl (1707-1778), renowned Swedish botanist
The Species Plantarum (1753) by Linnaeus is considered as the starting point of botanical nomenclature for seed plants. He introduced the so-called binomial nomenclaturé maintaining that the name of a species should be a binary combination, consisting of the name of the genus followed by a single epithet, e.g. *Pelargonium triste.*
Twenty species of *Pelargonium* were described in the Species Plantarum, but all of them were placed under the genus *Geranium,* as Linnaeus did not distinguish the genus *Pelargonium.* Linnaeus referred to his own botanical works and quoted from works by Hermann (1687 & 1689), Boerhaave (1720), Dillenius (1732), Burman (1738) and Van Royen (1740).

Burman, Nicolaas Laurens (1733-1793), Dutch physician and botanist, son of Johannes Burman
Burman (filius) described several new species of *Pelargonium* in his "Specimen botanicum de Geraniis" published in August 1759. All the species were placed under the genus *Geranium* and some of them were illustrated.

Cavanilles, Antonio Jose (1745-1804), Spanish botanist
The Geraniaceae was published in volume 4 of Cavanilles' "Monadelphiae classis dissertationes decem" (1787). He divided the genus *Geranium* into one section with actinomorphic flowers and another section with zygomorphic flowers. All the *Pelargonium* species were placed in the section with zygomorphic flowers, further subdividing the section on the basis of leaf characters. Seventy-one species of *Pelargonium* were described and illustrated in this work.

L'Heritier de Brutelle, Charles-Louis (1746-1800), French magistrate and botanist
L'Heritier was the first to use the name *Pelargonium* after the starting point of the botanical nomenclature, and is therefore considered as the author of the genus. His unfinished manuscript "Compendium Generalogium" dated 1789, was never published and after his death it was acquired by A.P. de Candolle. Proofsheets of this work are at present at the "Conservatoire Botanique", Geneva. Eighty-nine species of *Pelargonium* were described in this manuscript, and these descriptions have been copied literally in the first edition of Aiton's Hortus Kewensis, published in 1789. For this reason I am convinced that L'Heritier should be cited as the author of these species. L'Heritier also published his "Geraniologia" in 1792. This work consists of 44 uncoloured engravings of which most were redrawn from drawings by P.J. Redoute.

Thunberg, Carl Peter (1743-1828), Swedish botanist and explorer
Thunberg was a student of Linnaeus at Uppsala. He visited the Cape from 1772 to 1775 and has been called the "father of South African botany". The knowledge of the plants as well as the plant material he acquired on his extensive collecting trips, enabled him to write his "Prodromus Plantarum Capensium" (1794-1800). The Geraniaceae was included in volume one of this work in 1794. Thunberg started on a Flora Capensis in 1807 but he never finished the work. Thunberg did not recognize the genus *Pelargonium* and all the species were placed under *Geranium.*

Jacquin, Nikolaus Joseph Von (1727-1817), Austrian botanist
Jacquin worked as a botanist and botanical artist before he became a professor at Vienna. His knowledge of the Cape flora was largely derived from collections of Francis Boos and George Scholl, who were sent to the Cape by Emperor Joseph II in 1785 to collect African plants for the Imperial Garden at Schonbrünn.
Included in Jacquin's "Icones plantarum rariorum" (1781-1795), "Collectanea" (1786-1796) and "Plantarum rariorum horti caesarei schoenbrunnensis" (1797-1804), are descriptions and beautiful copper engravings of a large number of *Pelargonium* species.

Andrews, Henry C. (1799-1830), English botanical artist and engraver
Several natural species of *Pelargonium* as well as horticultural hybrids were described in the ten volumes of Andrews' "Botanists repository" published in 1797-1815. The hand-coloured engraved plates were made by Andrews personally, but the text was written by Kennedy, Haworth and Jackson.
The "Geraniums" by Andrews is a two volume work containing hand drawn and coloured figures of all the known natural species and numerous beautiful hybrids: They were described, engraved, drawn and coloured from cultivated plants.

Sweet, Robert (1783-1835), British horticulturist

The five volumes of Sweet's "Geraniaceae" were published in London between 1820 and 1830. Sweet distinguished the genera *Geranium, Erodium, Monsonia, Sarcocaulon* and *Pelargonium*, but some of the sections of *Pelargonium* like Campylia, Hoarea, Jenkinsonia etc., were raised to the status of genera.

Natural species and hybrids cultivated in the gardens of Great Britain were illustrated in colour and described with directions for their treatment. The hand-coloured copper engravings were based on drawings by E.D. Smith and a few by M. Hart.

Candolle, Augustin Pyramus de (1778-1841), Swiss botanist

De Candolle was professor of Botany in Geneva where his family had for generations enjoyed a great reputation. He divided the genus *Pelargonium* into 12 sections in volume 1 of his "Prodromus systematis naturalis regni vegetabilis" published in 1824. He compiled short descriptions of 369 natural species and hybrids, but they were not illustrated.

Ecklon, Christian Friederich (1795-1868) and Zeyher, Karl Ludwig (1799-1858), German botanists and travellers

Zeyher came to the Cape as a professional collector in 1822 and stayed intermittently for about 25 years. Ecklon who was also a professional collector, came one year later. The few years during which these two Germans were together in South Africa, led to the publication of their "Enumeratio plantarum Africae Australis". The Geraniaceae is dealt with in volume one and was published in 1835. The sections of *Pelargonium* were raised to the generic status and a large number of new species was described. Most of their new species were in fact synonyms of existing ones.

Harvey, William Henry (1811-1866), Irish botanist and traveller

Harvey was the Colonial Treasurer at the Cape from 1835 to 1842. In conjunction with O. W. Sonder he produced the first volume of "Flora Capensis" in 1860 of which the Geraniaceae was the compilation by Harvey himself. He divided *Pelargonium* into the following 15 sections (the number of species in each section is bracketed): Hoarea (42), Seymouria (4), Polyactium (24), Otidia (6), Ligularia (20), Jenkinsonia (3), Myrrhidium (6), Peristera (7), Campylia (8), Dibrachya (2), Eumorpha (7), Glaucophyllum (5), Ciconium (4), Cortusina (7) and Pelargium (22). The diagnostic features for the divisions were based on habit, leaf and flower characters. Harvey recognized a total number of 163 natural species of *Pelargonium*.

Knuth, Reinhard (1874-1957), German botanist

Knuth was a student of Engler and became professor in Berlin. He attained his doctorate degree with a dissertation on the genus *Geranium* and afterwards furthered his study on the Geraniaceae. His monumental work on the Geraniaceae was published in "Das Pflanzenreich" of Engler in 1912. Knuth divided the Geraniaceae into the tribes Geranieae, Biebersteinieae, Wendtieae, Vivianeae and Dirachmeae. The tribe Geranieae consists of the five genera *Geranium, Erodium, Monsonia, Sarcocaulon* and *Pelargonium*.

Knuth recognized the same 15 sections of *Pelargonium* as Harvey (1860) did, but he included more species in each section. He submitted keys to the identification of the sections and species.

A short summary of the sections and their diagnostic features are given below. The number of species in each section as recognized by Knuth is bracketed, and a list is compiled of those species described and illustrated in this work.

1. **Section Hoarea** DC. (48 species), named in honour of Sir Richard Colt Hoare. Stemless perennials with 1-3 but usually a single tuber which is often turnip-shaped. The flowers usually have five petals and five fertile stamens.
   The section contains beautiful species but they are difficult to propagate and dormant for a part of the year.

   Pelargonium hirsutum var. *melananthum*
   P. longifolium
   P. oblongatum
   P. pinnatum
   P. rapaceum

2. **Section Seymouria** (Sweet) Harv. (5 species)
   The characters of this section are identical with those of the section Hoarea except that the flowers have only two petals.
   P. asarifolium

3. **Section Polyactium** (Eckl. & Zeyh.) DC. (27 species)
   Polyactium refers to the star-like, many-flowered inflorescences of most of the species in this section. The plants usually have a subterranean tuber. Their leaves are lobed or pinnately divided. The flowers have subequal petals and a long spur and they are usually night-scented.

   P. fulgidum
   P. gibbosum
   P. lobatum
   P. luridum
   P. pulverulentum
   P. schizopetalum
   P. triste

4. **Section Otidia** (Lindl.) Harv. (8 species)
   The members of this section have thick and succulent stems which are often stout and knobby. Their leaves are fleshy and pinnately compound. The flowers have subequal petals and the upper petals are eared at their base, a characteristic which is denoted by the name Otidia. Five fertile stamens are present.

   P. carnosum
   P. crithmifolium

5. **Section Ligularia** (Sweet) Harv. (23 species)
   Species of this section have branched or much branched, subsucculent stems which are usually woody at the base. Their leaves are rarely entire being usually pinnately divided. The petals are subequal and spathulate with a tapering base. Seven fertile stamens are present.

   P. abrotanifolium
   P. crassipes
   P. fragile
   P. hirtum
   P. pulchellum

6. **Section Jenkinsonia** (Sweet) Harv. (4 species), named in honour of Jenkinson, an early horticulturist. Plants belonging to this section are half-shrubs with stems which are woody at the base or succulent. The leaves are palmately lobed and the flowers have four or five petals, the upper two petals being much larger

than the lower ones. Seven fertile stamens are present.

P. praemorsum
P. tetragonum

7. **Section Myrrhidium** DC. (9 species)
This section includes both perennial half-shrubs with diffuse and slender stems, and herbaceous annuals. Their leaves are pinnately divided. The flowers have four or five petals, the upper two petals being much larger than the lower ones. The sepals are membranous and prominently veined. Five or seven fertile stamens are present.

P. canariense
P. myrrhifolium var. betonicum
P. myrrhifolium var. fruticosum
P. urbanum var. bipinnatifidum

8. **Section Peristera** DC. (21 species)
Most species of the section Peristera are diffuse and herbaceous annuals, perennials are rare. The plants resemble *Erodium* and *Geranium* in habit. Their leaves are palmately lobed or pinnately divided and the flowers relatively small with the petals scarcely longer than the sepals. Five fertile stamens are present.

P. grossularioides

9. **Section Campylia** (Sweet) ·DC. (7 species)
All the species belonging to this section are small with short and erect stems. The entire, ovate, oblong, lanceolate or linear leaves are borne on long petioles with membranous stipules at their base. The upper two petals are broadly obovate and the three lower ones narrower. Five fertile stamens are present.

P. elegans
P. ovale
P. violareum

10. **Section Dibrachya** (Sweet) Harv. (4 species)
Species of this section are characterized by weak, thin and slender stems, mostly supported by other plants. The leaves are fleshy and peltate or cordate. The five obovate petals are unequal. Seven fertile stamens are present, the upper two being very short.

P. peltatum

11. **Section Eumorpha** (Eckl. & Zeyh.) Harv. (14 species)
Members of the section Eumorpha are shrubby and herbaceous. Their entire, palmately 5-7-veined and lobed or incised leaves are borne on long petioles. The five petals are unequal, the upper two being broader than the lower three. Seven fertile stamens are present.

P. alchemilloides
P. grandiflorum
P. tabulare

12. **Section Glaucophyllum** Harv. (6 species)
This section consists of half-shrubs. Their leaves are fleshy and simple or ternately lobed with the lamina articulated to the petiole. The petioles are often persistent and spiny. The five petals are unequal. Seven fertile stamens are present.

P. glaucum

13. **Section Ciconium** (Sweet) Harv. (4 species)
Species of this section are half-shrubs or shrubs with thick and fleshy stems. Their simple leaves are obovate or cordate-reniform and palmately veined. The five petals are all of the same colour. Seven fertile stamens are present, the upper two being very short. Among these species are the ancestors of the well-known zonal cultivars.

P. acetosum
P. inquinans
P. zonale

14. **Section Cortusina** DC. (19 species)
The section consists of half-shrubs with short, thick and fleshy stems which are often armed with persistent stipules. Their simple leaves are reniform or cordate, palmately lobed, silky or woolly and are borne on long petioles. The five petals are subequal, the upper two being broader than the lower ones. Six or seven fertile stamens are present.

P. echinatum
P. odoratissimum
P. reniforme
P. rhodanthum

15. **Section Pelargium** (DC.) Harv. (28 species)
This is a large section of shrubs and half-shrubs with woody and branched or very much branched stems. Their leaves are rarely entire and frequently lobed or divided. The upper two petals are longer and broader than the lower three. Seven fertile stamens are present.
Many of the species are ancestors of garden cultivars with scented leaves.

P. betulinum
P. capitatum
P. cordifolium
P. cucullatum
P. papilionaceum
P. scabrum
P. radens
P. vitifolium

The revision of *Pelargonium* by Knuth (1912) represents the most comprehensive and recent taxonomic study on the genus. This revision has its merits but it is, however, not without shortcomings. In time it has become clear that many of the species recognized by Knuth, are in fact synonyms of previously described species. The identification keys prove to be unsatisfactory, and some species were erroneously placed into sections. It is also doubtful whether the division of species into sections reveals their natural relationships.

A number of new species of *Pelargonium* have been described since 1912. Contributions to the knowledge of the genus were made by various authors such as Adamson & Salter (1950), Moore (1955), Müller (1963), Carolin (1961), Merxmüller & Schreiber (1966), Clifford (1970), Mastalerz (1971) and Webb (1971 & 1972).

A group of botanists from the Department of Botany at the University of Stellenbosch, is working under my direction on an all-inclusive taxonomic project of *Pelargonium*. It will stretch over a period of years and will, as far as possible, include all aspects of morphology, anatomy, cytology, palynology and chemotaxonomy of all the natural taxa of the genus. It has been suggested by many authors that extensive field work should be the prime requisite in order to revise the genus taxonomically. We are in the most fortunate position to be able to do such field work. The important features of the genus *Pelargonium* as known up to date, may be summarised as follows:

PELARGONIUM L'Herit. in Ait. Hort. Kew. ed. 1,2: 417 (1789).

Perennial shrubs, shrublets, acaulescent geophytes, rarely scramblers or annual herbs. *Stems* erect or decumbent, often woody at the base, soft-wooded or subsucculent, occasionally succulent, often viscid and aromatic, variously hairy, often glandular, sometimes with persistent spine-like stipules or petioles. *Leaves* usually petiolate, stipulate, alternate or opposite, entire to much dissected (even on same plant) or compound, variously hairy, often glandular and aromatic. *Inflorescence* a 2-many-flowered pseudo-umbel (oldest flowers occupying the centre), bracts present. *Flowers* rarely solitary, irregular (zygomorphic), 5-merous. *Sepals* imbricate, connate at base, receptacle forming a hypanthium with a nectariferous spur opening at base of posterior sepal, lower end of spur thickened and with a nectariferous gland. *Petals* usually 5, rarely 4, seldom 2, in 2 groups of 2 upper and 0-3 lower, imbricate, unguiculate (clawed) or sessile, rarely lacerate, usually variously coloured. *Disc* or exstaminal glands absent. *Stamens* 10, connate at base, 2-7 filaments bearing anthers (fertile), remaining ones often vestigial (staminodes), anthers dorsifixed. *Ovary* 5-lobed, 5-locular (rarely 3-4-locular by abortion), beaked (rostrate), with 2 ovules in each locule, hirsute; style of varying length; stigmas 5, usually filiform and reflexed. *Fruit* a rostrate schizocarp; mericarps rostrate, 1-seeded, tapering from the apex to the base and ending in a spirally twisted awn when ripe, with long hairs. *Seeds* ± oblong-obovoid; endosperm absent; embryo curved.

Pelargonium abrotanifolium

(Southernwood-leaved pelargonium)

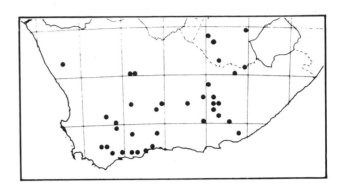

Artemesia abrotanum (Compositae), botanical name of Southernwood; folium (Latin), leaf; the leaves of this pelargonium resemble those of Southernwood.

Twiggy shrublet, branches slender.
Leaves grey-green, linearly divided, fragrant.
Flowers white, pink or mauve.

P. abrotanifolium (L.f.) Jacq. in Hort. Schoenbr. 2: 6, t.136 (1800); R. Knuth in Pflanzenr. 4, 129: 384 (1912).
Originally described by Linnaeus (filius) as *Geranium abrotanifolium* in L.f., Suppl.: 304 (1781).

Description: An aromatic and much-branched shrublet attaining a height of 1 m, though usually smaller. The slender, woody branches are glandular and often covered with the remains of petiole bases.
The feathery grey-green (canescent) leaves are pleasantly fragrant, due to a volatile substance exuded by the numerous glandular hairs. Leaf blades are 5-17 mm long and 5-19 mm broad, and vary considerably in structure. They are usually three or five segmented, each segment again being subdivided into three or more linear lobes. These lobes are channelled on their upper surface along the midribs. Small, lanceolate stipules are found at the base of the relatively long petioles.
Inflorescences are unbranched with few flowers. Each peduncle bears 1-5 flowers; they have five spathulate or obovate petals, of which the colour varies from white to pink or mauve, veined in red or purple. Seven of the ten filaments bear anthers. The flowering period extends throughout the greater part of the year.

Distribution: This species is widely distributed in the Cape Province and it occurs in the Orange Free State as well.
It grows in relatively arid habitats and is often found on rocky outcrops.

Remarks: According to Moore in Mastalerz (1971), *P. abrotanifolium* is the only species of the section Ligularia which is in general cultivation in the United States of America. The cultivation of this species was practised in Britain as early as 1791.
Cuttings root readily.

X1

X2

X3

Pelargonium acetosum

(Sorrel-leaved pelargonium)

Acetosus (Latin), sour; refers to the acid taste of the leaves.

Bushy shrublet, stems subsucculent.
Leaves obovate, cuneate, fleshy, margin red.
Flowers usually salmon-pink.

P.acetosum (L.) L'Herit. in Ait. Hort. Kew. ed. 1,2: 430 (1789); Knuth in Pflanzenr. 4,129: 440 (1912).
Originally described by Linnaeus as *Geranium acetosum* in Sp. Pl. ed. 1,2: 678 (1753).

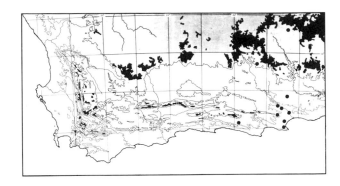

Description: A bushy, much-branched shrublet up to 60 cm tall. The branches are sparsely leaved, slender, smooth and rather succulent when young.
The glabrous leaves have an obovate outline with a cuneate base. They are glaucous-green, somewhat fleshy, 1-6,5 cm long and 1-5 cm broad. The red and coarsely crenate margin of the leaves is a characteristic fea-ture of the species. Obliquely ovate stipules are found at the base of the relatively short petioles.
The inflorescences are unbranched and the peduncles bear 2-7-flowered pseudo-umbels. Long glandular hairs occur on the pedicel, hypanthium and calyx. The colour of the attractive flowers shades from brilliant salmon-pink to almost white. The petals are relatively long and narrow, the upper two being typically erect. Five fertile stamens are present. The flowering period extends over the whole year.

Distribution: This species is confined to the Eastern Cape. It occurs from the Conga River in the west to Kirkwood in the east, and as far north as Steynsburg. It grows on stony hillsides and in dry grasslands.

Remarks: *P. acetosum* would be an acquisition for cultivation in gardens in drier parts of South Africa. Cuttings root readily.
It was first cultivated in the Chelsea Garden, England, in 1724.

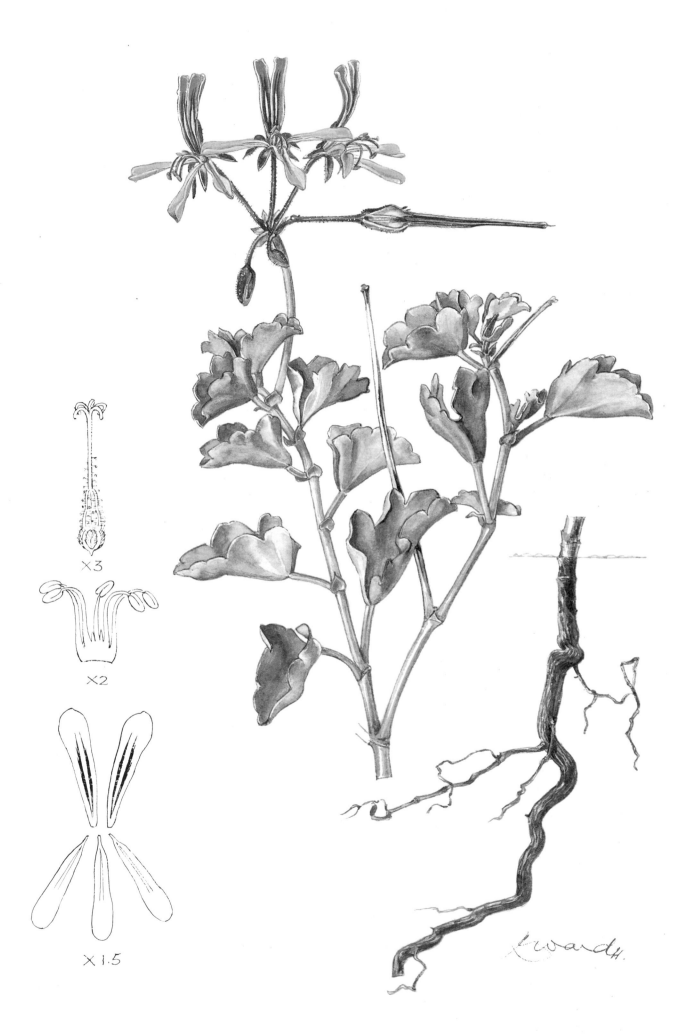

×3

×2

×1.5

Lwardₕ.

Pelargonium alchemilloides

Lady's Mantle-leaved pelargonium; "Wilde malva").

Alchemilla (member of the Rosaceae or rose family, commonly known as Lady's Mantle). -oides (Greek), resemblance.

Decumbent herb with straggling stems. Leaves palmately lobed, silky, stipules broadly ovate.
Flowers white to deep pink.

P. alchemilloides (L.) L'Herit. in Ait. Hort. Kew. ed. 1,2:419 (1789); Knuth in Pflanzenr. 4, 129: 428 (1912).
Originally described by Linnaeus as *Geranium alchemilloides* in Sp. Pl. ed. 1,2:678 (1753).

Synonyms: *P. multibracteatum, P. usambarense, P. malvaefolium, P. alchemillifolium,* etc.

Description: A decumbent perennial herb usually ca. 20 cm tall with a woody stoloniferous to tuberous rootstock. The stems are herbaceous, slender, often straggling, covered with long, rather coarse hairs which are visible to the naked eye.
The attractive leaves which have an almost silky appearance, due to appressed hairs, are mostly roundish in outline with a cordate base They are 2-7 cm in diameter, usually 5-palmately lobed and the margins of the

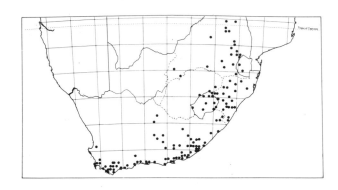

segments are irregularly toothed. A purple, brown or even dark red, horse-shoe-shaped zonal marking is characteristic of many leaves. Ovate or broadly ovate stipules are found at the base of the long petioles, a feature which distinguishes this species from the very similar *P. tabulare* which has lanceolate stipules.
The inflorescence is umbel-like with 3-6- but sometimes up to 15 flowers. Plants growing in the South Western Cape usually have white or cream flowers occasionally with pink or red markings, but those occurring in the Eastern Cape and other Provinces vary from pale to deep pink. Seven fertile stamens are present. It flowers in Spring but flowers may be found almost throughout the year.

Distribution: This is probably the most widely distributed pelargonium in Africa, occurring in all four provinces of South Africa, in Lesotho, Swaziland, Rhodesia, Tanzania, Kenya, Ethiopia and Somalia. It is often found in disturbed areas.

Remarks: The large number of synonyms resulted from the variable leaf characters as well as from the wide distribution of the species Kokwaro (1969) distinguishes two subspecies, of which only subspecies *alchemilloides,* is represented in South Africa. A decoction of the root is used in Lesotho to bathe a feverish patient. Reported to be cultivated by J. Bobart in Britain as early as 1693.

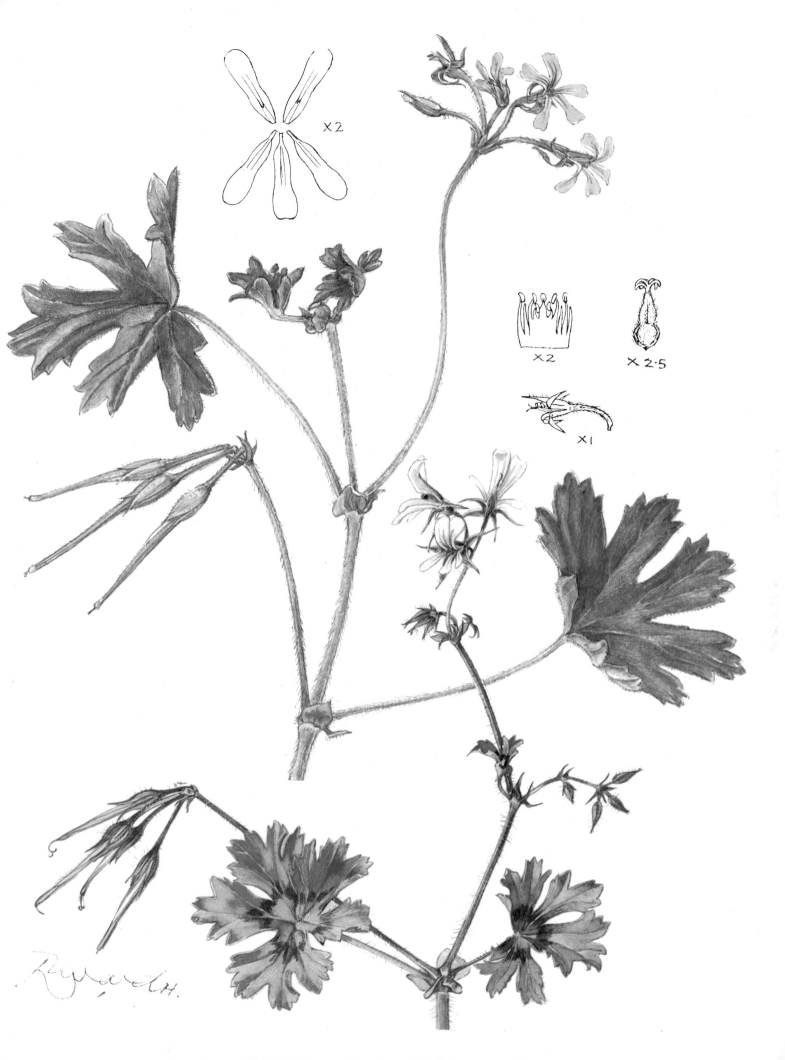

×2

×2

×2·5

×1

Pelargonium asarifolium

(Asarum-leaved pelargonium)

Asarum is a genus of the family Aristolochiaceae; -folium (Latin), leaf; resembling the leaf of *Asarum*.

Small stemless geophyte.
Leaves cordate, glabrous above, tomentose underneath.
Flowers with only two reflexed petals, dark red.

P. asarifolium (Sweet) G. Don in Gen. Syst. 1: 731 (1831); Knuth in Pflanzenr. 4, 129: 349 (1912).
Originally described as *Seymouria asarifolium* by Sweet in his Geraniaceae 3: t. 206 (1824).

Description: A small and stemless geophyte with an elongated tuber which can reach a length of up to 7 cm.
The entire, cordate or roundish-cordate leaves, are borne on hairy petioles. Their leaf blades have entire margins. The upper surface of the leaves is glabrous and shiny green, while the under surface is grey due to the thick matted cover of hairs (tomentose).

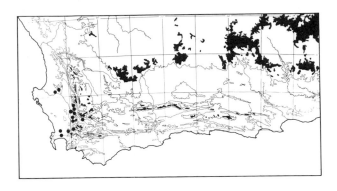

Most leaves are 4-5 cm long and 3-4 cm broad. Small and sharp-pointed stipules are found at the broad base of the petioles.
When the leaves start to wither in the hot summer months, long and villous scapes appear from the rootstock. These scapes are usually branched to form umbel-like inflorescences, each bearing 7-12 flowers. The flowers are peculiar and fascinating in having only two petals which are sharply reflexed from about their middle. A long staminal column with the same dark red or dark purple colour as the two petals, compensates for the absence of the lower petals. Five fertile stamens are present. Plants flower from December to June.

Distribution: This interesting species is apparently confined to a relatively small area in the South Western Cape. It has been recorded from the districts of Piketberg, Tulbagh, Worcester and Stellenbosch only. The plants seem to grow equally well in stony, sand or even clayey soils.

Remarks: The drawing accompanying the original description of the species, was made from a plant which flowered in the nursery of Colvill (England) in 1822. Sweet proposed it as a distinct genus Seymouria, in honour of Mrs. Seymour (Bedfordshire). She was a great admirer of Alpine plants, to which this species bears a strong resemblance.

×2.5

×3

×3

×3

×1

Pelargonium betulinum

(Birch-leaf pelargonium; "Kanferblaar"; "Maagpynbossie"; "Suurbos")

The leaves resemble those of the well-known European Birch trees (*Betula* spp.), hence the specific epithet *betulinum*.

Erect or sprawling shrub or bush.
Leaves asymmetrical oval or ovate.
Flowers large, pink to purplish.

P. betulinum (L.) L'Herit. in Ait. Hort. Kew. ed. 1,2: 429 (1789); Knuth in Pflanzenr. 4,129: 457 (1912).
Originally described by Linnaeus as *Geranium betulinum* in Sp. Pl. ed. 1,2: 679 (1753).

Description: A small erect or sprawling shrub or bush, 0,3-1,3 m high with rather woody branches which are glabrous or minutely downy.
The leaves are borne on petioles which are normally ca. 1 cm long. The asymmetrical oval or ovate leaf blades are 1-3 cm long and 0,7-2,5 cm broad. They are rather hard and almost glabrous or covered with fine short hairs. The margins of the leaves are dentated, the teeth being unequal and often red-tipped. Plants collected from the Mossel Bay - Knysna areas show a much coarser leaf margin than those from the South Western Cape.

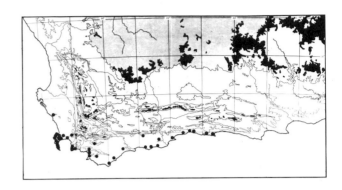

The plants are very conspicuous in the veld when in full bloom due to the large and attractive flowers. The inflorescence is umbel-like and 1-6 but usually 3-4 flowered. Pink and purplish flowers with darker streaks are common, but whitish flowers with dark streaks are also found. The upper two petals are broader and darker coloured than the three lower ones. Seven fertile stamens with orange-coloured anthers are present. This species flowers from August to October, although odd flowers occur as late as February.

Distribution: *P. betulinum* occurs in the South Western Cape and South Cape and is almost confined to coastal areas. It has been recorded from Ysterfontein to Knysna. The plants usually grow on sandy dunes or flats.

Remarks: This is one of the natural species which deserves a place in the garden. It is readily propagated by cuttings and was introduced to Britain by Francis Masson in 1786. Masson was an English gardener at Kew who was sent to the Cape by Sir Joseph Banks to collect plants for Kew Gardens.
The plants were used medicinally for coughs and other chest troubles, the vapour from steamed leaves being inhaled. The vernacular name "Kanferblaar" is derived from the camphor-like odour of crushed leaves.
P. betulinum is often found growing in association with *P. cucullatum,* and hybrids of these two species are not uncommon.

X2

XL

Pelargonium canariense

Canariensis (Latin), referring to the Canary Islands.

Halfshrub, branches hairy.
Leaves 3-lobed, silky, tending to grey.
Flowers pale pink, four petals, five fertile stamens.

P. canariense Willd. in Hort. Berol. 1,2: t.17 (1804) — original description.

Synonym: *P. candicans* Spreng. in Syst. Veg. 3: 57 (1826); Knuth in Pflanzenr. 4, 129: 399 (1912).

Description: An attractive, low-growing halfshrub attaining a height of ca. 30 cm. The procumbent branches are covered with long soft hairs (villous).
The leaves are ca. 3 cm long and 2,5 cm broad, elongate cordate in outline and more

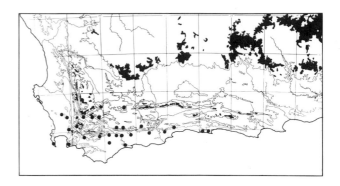

or less 3-lobed with the terminal lobe usually much larger than the other two. Their margin is crenulate, they tend to grey (canescent) and are silky on both sides. Puberulous, ovate to broadly ovate stipules are found at the base of the very long and hairy petioles.

The umbel-like inflorescences bear 1-4 flowers each. This species has only four, white to pale pink petals per flower. The two larger upper petals are decorated with beetroot-red veins, and the lower two with broad mauve strips. Five filaments bear anthers. Flowers can be found throughout the year.

Distribution: It occurs from Gydouw Pass (Ceres district) in the north west, to Plettenberg Bay in the south east.

Remarks: This pelargonium is commonly known as *P. candicans.* The specific epithet *canariense,* however, has priority and is therefore the legitimate name.
When Willdenov originally described it, he was under the impression that the plant came from the Canary Islands — hence the misleading name.

6

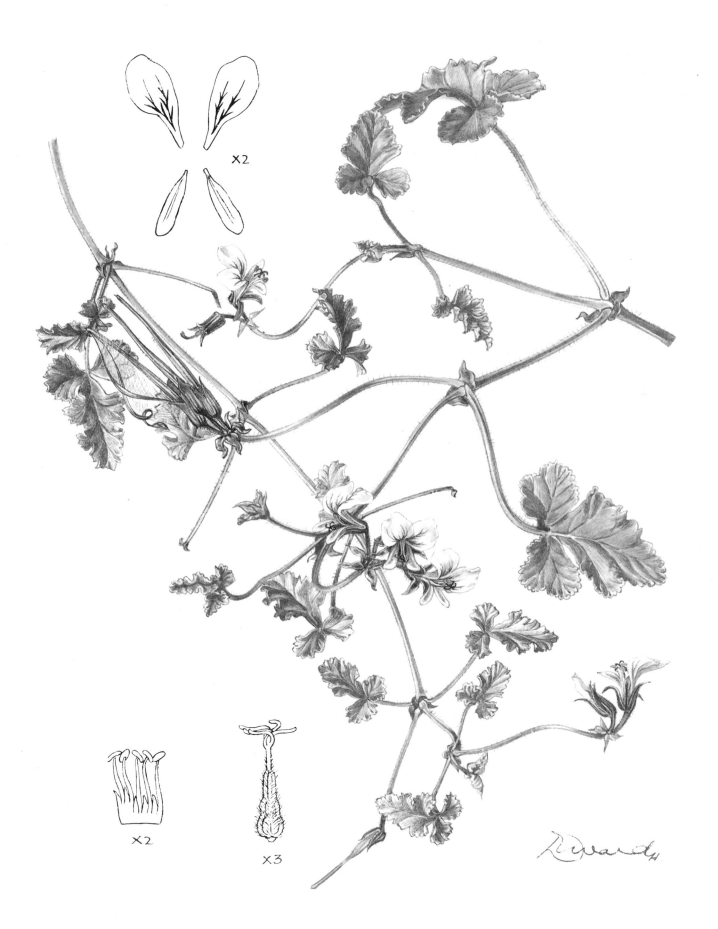

X2

X2

X3

Pelargonium capitatum

(Rose-scented pelargonium)

Capitatus (Latin), refers to the head-like (capitate) inflorescence.

Low growing shrublet or bush with sprawling branches.
Leaves 3-6-lobed, velvety, crinkled, aromatic.
Flowers borne in many-flowered, compact heads.

P. capitatum (L.) L'Herit. in Ait. Hort. Kew. ed. 1,2: 425 (1789); Knuth in Pflanzenr. 4,129: 467 (1912).
Originally described by Linnaeus as *Geranium capitatum* in Sp. Pl. ed. 1,2: 678 (1753).

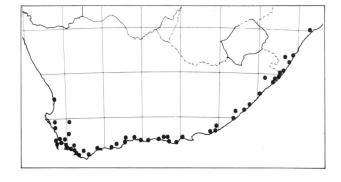

Description: A shrubby or bushy and low growing plant, 0,25-1 m tall and up to 1,6 m in diameter. The sprawling or erect stems are soft wooded. Individual side-branches can attain a length of 60 cm.
Stems and leaves are sweetly scented when bruised, and covered with long soft hairs of variable density (villous to densely villous). The crinkled, velvety leaves with a cordate base, are shallowly to deeply 3-6-lobed and generally ca. 4,5 cm long and ca. 6 cm broad. The segments themselves may also be lobed with the leaf margin toothed throughout. Leaves of the plants in the Eastern Cape and Natal, tend to be larger than those of the South Western Cape.
The inflorescences are capitate and 8-20-flowered. Due to the very short pedicels of the flowers and the large number of flowers, the heads appear rather compact. The peduncles, bracts and calyx are also villous. The common colour of the petals is cyclamen-purple with beetroot-purple stripes on the two upper and slightly larger petals. Pale pink and dark pink-purple flowers also occur. Seven fertile stamens are present. Plants start to flower in the early Spring, although they may be found in flower almost throughout the year.

Distribution: This species occurs from Lambert's Bay all along the coast through the Transkei to Zululand. It is abundant on sand dunes or low hill-sides near the sea. As in the case of other pelargoniums, it is commonly found growing in disturbed areas.

Remarks: This is one of the pelargoniums which is cultivated for oil of geranium. It grows very easily from cuttings and was introduced to Britain by Bentick in 1690.
P. capitatum is closely allied to *P. vitifolium.*

X1.

X1.

X2

Pelargonium carnosum

(Fleshy-stalked pelargonium)

Carnosus (Latin), fleshy or succulent; refers to the succulent stems.

Succulent shrub with smooth stems.
Leaves pinnately divided, often somewhat fleshy.
Flowers small, white to greenish-yellow, five fertile stamens.

P. carnosum (L.) L'Herit. in Ait. Hort. Kew. ed. 1,2: 421 (1789); Knuth in Pflanzenr. 4, 129: 370 (1912).
Originally described by Linnaeus as *Geranium carnosum* in Cent. Pl. 1: 281 (1755).

Description: A succulent and branched shrub with a height of 0,3-1,2 m. The thick and fleshy stems have a knobby appearance due to the somewhat swollen nodes. Young branches are sparsely covered with whitish hairs, but they become smooth with age.
This species shows great variability in leaf characters. The leaves are usually pinnately divided (bipinnati-partite), channelled along the midrib, and variably fleshy. Their pubes-

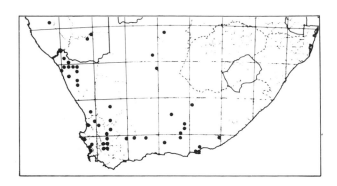

cence may vary from sparse to dense puberulous. The leaf blades are 3-17 cm long and 2-5 cm broad and the length of the petioles varies from 1-10 cm. Small triangular and membranous stipules are present at the thickened base of the petioles.
The inflorescence varies considerably on plants from different habitats. Peduncles are usually branched to bear several pseudo-umbels with 2-8 flowers each. The colour of the relatively small flowers varies from white to greenish-yellow with reddish streaks on the two, slightly larger, upper petals. Only five fertile stamens are present. The plants flower from January to March.

Distribution: *P. carnosum* occurs in the southern parts of South West Africa, Richtersveld, Namaqualand, Karoo, South Western Cape and dry areas in the Eastern Province.
It is confined to xerophytic habitats and grows in sandy soil as a component of semi-desert and karroid vegetation.

Remarks: In the original description of *P. carnosum,* Linnaeus referred to an illustration and description by Dillenius (1732). This illustration perfectly matches the plants which are generally known as *P. ferulaceum.* It has been suggested by Dyer (1953) that *P. carnosum* and *P. ferulaceum* are con-specific, a view which is held in this work.

x2   x3   x3   x1.

Pelargonium cordifolium

(Heart-leaved pelargonium)

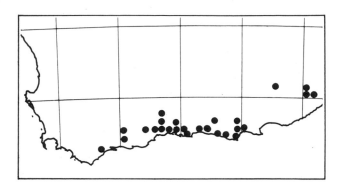

Cordatus (Latin), cordate or heart-shaped; -folium (Latin), leaf; refers to the shape of the leaves.

Branched and aromatic shrub.
Leaves cordate, pale and velvety beneath.
Flowers pink to purple, upper petals much larger than lower ones.

P. cordifolium (Cav.) Curt. in Bot. Mag. 5: t. 165 (1791).
Originally described by Cavanilles as *Geranium cordifolium* in Cav. Diss. 4: 240, t. 117, f.3 (1787).

Synonym: *P. cordatum* L'Herit. in Ait. Hort. Kew. ed. 1,2: 427 (1789); Knuth in Pflanzenr. 4, 129: 464 (1912).

Description: A spreading and branched aromatic shrub reaching a height of more than 1,5 m but usually ca. 1 m tall. The main stem is woody at the base, while the side-branches are herbaceous and covered with hairs of variable density.
It has beautiful foliage, the leaves being distinctly cordate and flat, or curled. The leaves are usually ca. 6 cm long and ca. 5 cm broad, finely to coarsely toothed, and un-lobed or shallowly lobed. They are usually lighter in colour beneath with a matted, soft, wool-like hairiness giving them a velvety texture. The degree of hairiness varies considerably. Plants growing in the Humansdorp area have densely-haired leaves, while those of some plants in the George district, are almost glabrous with only very short hairs underneath.

The attractive flowers are borne in a branched inflorescence terminating in several 4-8-flowered umbel-like groups. The flowers are papilionaceous like those of *P. papilionaceum,* with the two upper petals much larger than the three lower ones. The upper petals are usually pink or purple with darker purple veins. The lower petals are lighter in colour. Seven fertile stamens are present. This species can be found in flower from June to January but it is usually in full bloom during Spring.

Distribution: Occurs mainly near the coast in the Southern and Eastern Cape from the Bredasdorp area eastwards to King Williamstown. It usually grows in rather moist places in Fynbos or at the margins of forests or even in forests.

Remarks: *P. cordifolium* is another species worth-while cultivating as cuttings strike root freely. It was introduced to England in the year 1774 by Masson. Three varieties have been recognized on basis of leaf characteristics.

X1

X2

Pelargonium crassipes

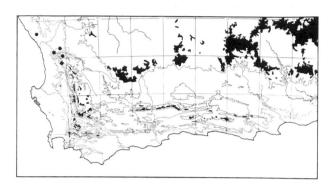

Crassus (Latin), thick; refers to the thickened stem.

Small shrublet with armed stems.
Leaves finely divided, bases of petioles
large and persistent.
Flowers relatively small, usually pink.

P. crassipes Harv. in Fl. Cap. 1: 281 (1860) — original description; Knuth in Pflanzenr. 4,129: 381 (1912).

Synonym: *P. caucalidifolium* Schltr. in Bot. Jahrb. 27: 150 (1900); Knuth in Pflanzenr. 4,129: 381 (1912).

Description: A small shrublet 15-30 cm tall with a rather fleshy and unbranched or sparsely branched main stem. The stem is thickly covered with persistent and hardened bases of old petioles, resembling the dry leaves of tree aloes.

The finely divided leaves have an oblong-ovate outline and they are 2,5-7 cm long and 1-2,5 cm broad. Leaves on the flowering stems are usually much smaller. All the leaves are densely covered with coarse hairs and numerous glands. The exceptionally large and tapering bases of the petioles form persistent spine-like structures which bend downwards with age. The narrow stipules are almost completely adnate to the bases of the petioles, their apices however remain free.

The thin flowering stems are usually branched. Relatively small flowers are borne in 2-10-flowered umbel-like inflorescences. The colour of the petals varies from pale pink to mauve with markings of a darker tone. Seven fertile stamens are present. The plants flower during the early Spring.

Distribution: This is another pelargonium which is apparently confined to a relatively small area of Namaqualand. It has been recorded as growing only between Lutzville (near Van Rhynsdorp) and Clanwilliam, a region which lies in the Succulent Karoo.

Remarks: Knuth (1912) considered *P. crassipes* and *P. caucalidifolium* as two different species, but it is almost certain that they are conspecific. *P. crassipes* is the earlier and consequently correct name of this species.

X1.

X2.

X2.   X3.

Edward H

Pelargonium crithmifolium

(Samphire-leaved pelargonium).

Crithmum is the generic name of Samphire (St. Peter's herb), -folium (Latin), leaf; the leaves resemble those of Samphire.

Succulent shrublet, stems thick and knobby.
Leaves fleshy, pinnately divided.
Panicled inflorescence becoming thorny with age.

P. crithmifolium J.E. Sm. in Ic. Pict. Pl. Rar. 1: 13, t.13 (1793) — original description.

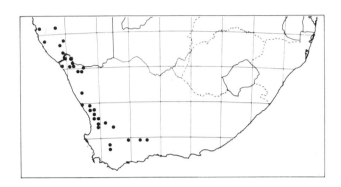

Description: A succulent and branched shrublet attaining a height of 0,5 m, though usually much smaller. The thick, fleshy and knobby stems have a green or yellowish-green colour and a smooth or peeling bark. The fleshy and smooth leaves are usually crowded on the upper part of the branches. They are pinnately divided (sub-bipinnatipartite or -fid), 5-12 cm long and 2-8 cm broad. The narrow lobes of the leaves have a dilated, irregularly toothed apex and a slightly pubescent margin. Leaf blades are borne on relatively long petioles which are channelled on the upper side. The stipules are small, ovate and membranous.

Large and very much branched, panicled inflorescences are borne on the top of the branches. These inflorescences become hard and thorny with age and they are persistent for at least one year. Several pseudo-umbels, each with 4-6 flowers are found on each inflorescence. The colour of the flowers is white with red markings near the base of the upper two petals. These upper two petals with a characteristically crisped base, are slightly larger than the lower three. Five filaments bear anthers. It flowers from May to October.

Distribution: This species occurs in the southern part of South West Africa, Richtersveld and Namaqualand. It grows in dry, hot habitats, and is often found in fissures of rocky outcrops.

Remarks: Many authors such as Knuth (1912), consider *P. paniculatum* as a synonym of *P. crithmifolium*. *P. paniculatum* is said to have an unbranched stem and a non-persistent inflorescence.
P. crithmifolium was cultivated in England as early as 1792.

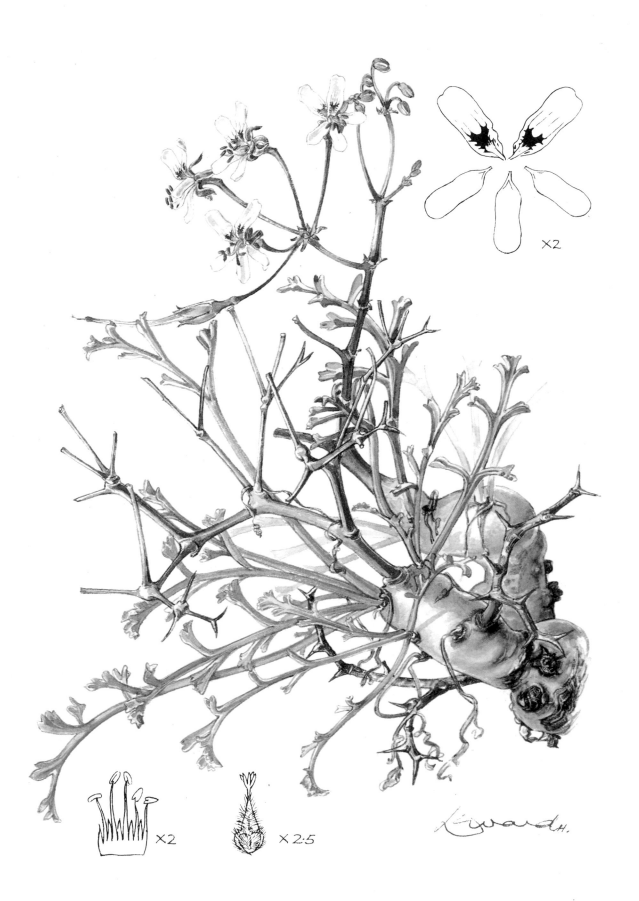

×2

×2   ×2·5

Pelargonium cucullatum

(Hooded-leaf pelargonium; Tree pelargonium; "Wilde malva")

Cucullatus (Latin), hooded; refers to the leaves which are typically cupped.

Relatively tall shrub.
Leaves kidney-shaped, cupped, villous.
Flowers large, pink to purple.

P. cucullatum (L.) L'Herit. in Ait. Hort. Kew. ed. 1,2: 426 (1789); Knuth in Pflanzenr. 4, 129: 466 (1912).
Originally described by Linnaeus as *Geranium cucullatum* in Sp. Pl. ed. 1,2: 677 (1753).

Description. One of the tallest shrub pelargoniums, attaining a height of more than 2 m. The main stem is up to 2 cm thick, half-woody at the base but the younger parts and side branches are more or less succulent.
The leaves of the Table Mountain form are usually ca. 4,5 cm long and ca. 6 cm broad, kidney-shaped, typically hooded and covered with long soft hairs (villous). The margins of the leaves are sometimes reddish, irregularly toothed but not prominently incised. Plants from other localities along the coast have less cupped leaves with angularly incised margins and fewer hairs. The latter type of leaf represents a transitional form to the harsh leaves of the closely allied *Pelargonium angulosum* which usually occurs further inland.
Inflorescences are umbel-like and 4-

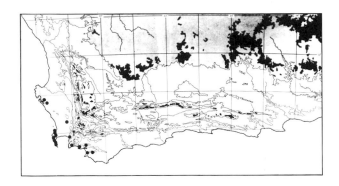

10-flowered. The colour of the relatively large flowers varies from pink-purple to dark purple, but white flowers also occur. The upper two petals with dark red veins, are slightly larger than the three lower ones. This species is in full bloom during Spring, but flowers can be found throughout the year. Orange-coloured pollen borne on seven fertile stamens, contributes to the beauty of the flowers.

Distribution: Restricted to the South Western Cape where it occurs near the coast between Saldanha and Elim (west of Bredasdorp). It is well represented in the Cape Peninsula.

Remarks: *P. cucullatum* is the most conspicuous pelargonium of the South Western Cape especially when growing in dense masses. The flowers are faintly scented and are reportedly visited by sunbirds, long-beaked flies, moths and butterflies.
It is the principal ancestor of the Regal pelargoniums and was introduced into cultivation in England by Bentick in 1690. In William Harvey's time at the Cape (1838-40) it was cultivated as an ornamental hedge-row plant in Cape Town and it is the first record of a pelargonium to be so used. The roots and leaves were formerly used as an astringent in cases of colic and diarrhoea, as an emmollient or an antispasmodic.
Natural hybrids occur between *P. cucullatum* and *P. betulinum, P. angulosum* and *P. saniculaefolium*.

×1·5

×2

×1

×1

R.Ward.
H.

Pelargonium echinatum

(Prickly-stemmed pelargonium)

Echinatus (Latin), echinate, i.e. armed with prickles or spines; refers to the spine-like stipules on the stem.

Perennial shrublet with fleshy and spiny stems.
Leaves cordate-ovate and 3-5-7-lobed.
Flowers white to pink to purple with darker blotches.

P. *echinatum* Curt. in Bot. Mag. 9: t.309 (1795) — original description; Knuth in Pflanzenr. 4, 129: 445 (1912).

Description: A perennial, erect shrublet with tuberous roots, reaching a height of 60 cm, although usually much smaller. The branched stem is fleshy and armed with persistent, recurved, spine-like stipules. Flowering stems are relatively thin and smooth or slightly downy.
The cordate-ovate leaves with crenate or crenulate margins are 3-5-7-lobed, 2-3 cm long and 3-4 cm broad. Leaves on branches of the main stem are much larger than those on the flowering stems. The under surfaces of the greyish-green leaf blades are lighter in

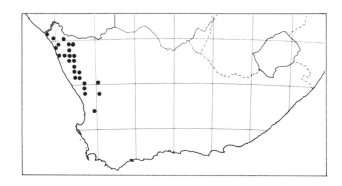

colour and also more hairy (tomentose) than the upper surfaces. Awl-shaped stipules are found at the base of the relatively long petioles.
The inflorescence consists of a relatively long peduncle terminating in a group of 3-8 flowers with an umbel-like appearance. The colour of the flowers varies from white to pink to brilliant purple. Darker blotches and markings are present on the upper two petals and occasionally darker blotches are also found on the lower three petals. One of the filaments is exceptionally broad and in some cases devoid of an anther, with the result that the number of fertile stamens varies from six to seven. Plants flower from July to November.

Distribution: This is one of the beautiful species of Namaqualand, occuring from Clanwilliam to the Richtersveld. It usually grows on dry, stony slopes under the protection of bushes or overhanging rocks.

Remarks: The succulent stem, tuberous roots and deciduous leaves are all adaptations of this pelargonium to resist drought during the hot and dry summer.
The original description, as well as the accompanying colour illustration of this species, was done from a living plant cultivated in the Chelsea Garden (England) of Colvill (Curtis, 1795).
Cuttings strike readily especially when taken at the growing stage.

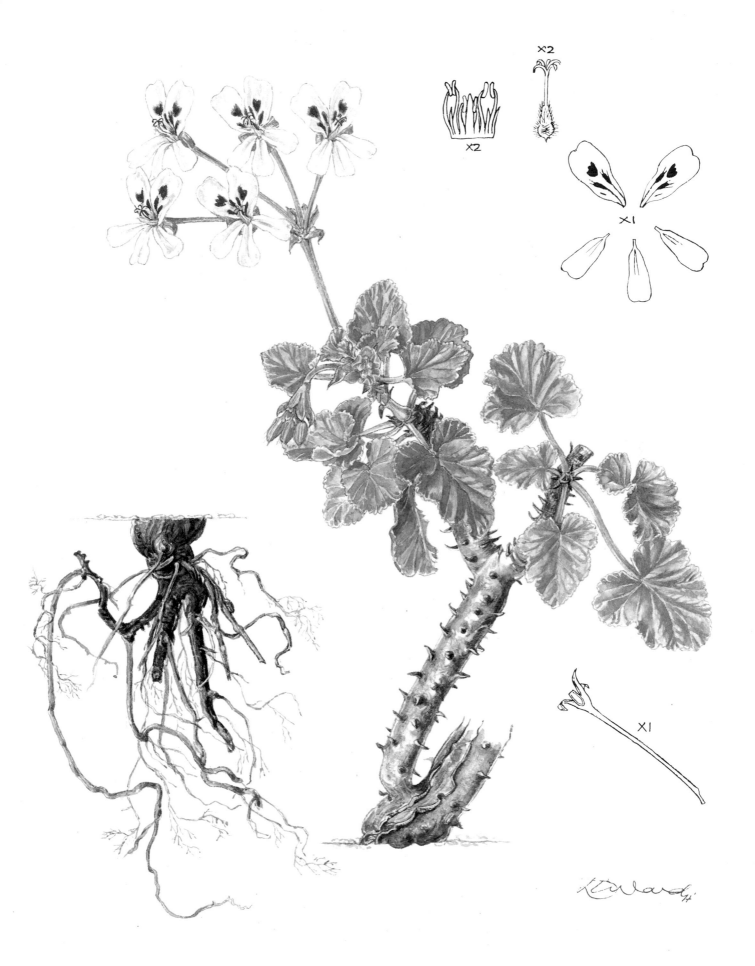

Pelargonium elegans

Elegans (Latin), elegant; apparently refers to the attractive appearance of the plants.

Tufted halfshrub, short stem with persistent stipules.
Leaves suborbicular, green, hard.
Flowers large, pale pink to lilac.

P. elegans (Andr.) Willd. in Sp. Pl. ed. 4,3: 655 (1800). Knuth in Pflanzenr. 4,129: 419 (1912).
Originally described by Andrews as *Geranium elegans* in Bot. Rep. 1: 28 (1798).

Description: A small, tufted halfshrub with a height of up to 25 cm and a short, usually unbranched stem. The stem is densely covered with the remains of old stipules.
The green, suborbicular leaves with a cordate base, have a relatively hard texture and coarsely serrate margin. Leaf blades are 2-4 cm in diameter and often slightly elongated. Mature leaves are almost glabrous with a few

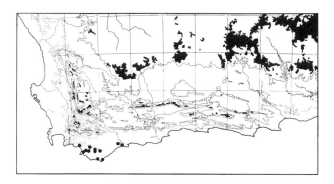

stiff, short hairs on the veins, and a ciliate margin. The petioles are relatively long (up to 10 cm). Large, deltoid and brick-coloured stipules are present.
Flowering branches usually give rise to three peduncles, each bearing 2-4 flowers. The colour of the large and attractive flowers varies from pale pink to lilac with dark purple veins on the two upper petals. Seven lilac-coloured filaments bear dark purple anthers. This species flowers from September to January.

Distribution: *P. elegans* is confined to coastal dunes and flats in the Hermanus, Gansbaai and Bredasdorp districts. It also occurs at Potberg near Malgas.

Remarks: This species is stoloniferous. The majority of plants of a population are connected up by subterranean stolons. Nevertheless, it is difficult to propagate artificially, as cuttings do not strike readily.
The flowers and leaves superficially resemble those of *P. betulinum* but the two species differ completely in habit.

X1

X1.5

X2

Pelargonium fragile

(Brittle-stalked pelargonium)

Fragilis (Latin), fragile; refers to the brittle stem.

Shrubby, stems slender and brittle.
Leaves deeply trifid or trifoliolate, margin irregularly incised.
Flowers large, cream-coloured with reddish stripes.

P. fragile (Andr.) Willd. in Sp. Pl. ed. 4,3: 686 (1800).
Originally described by Andrews as *Geranium fragile* in Andr. Bot. Rep. 1: t. 37 (1798).

Synonyms: *P. tripartitum* Willd. in Sp. Pl. ed. 4,3: 683 (1800); Knuth in Pflanzenr. 4,129: 388 (1912).

P. trifidum Jacq. in Hort. Schoenbr. 2: 5, t.134 (1797).

Description: A shrubby and aromatic plant, with several scrambling branches which may attain a height of 1 m when supported by other plants. The flexuose stems are slender, brittle and covered with soft hairs (pubescent).

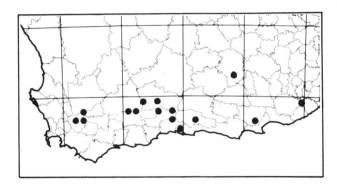

The rather fleshy leaves are variable in form and hairiness. They are normally 1,5-4 cm long and equally wide; deeply trifid or even trifoliolate with the margin of the segments /leaflets irregularly incised. The terminal segment/leaflet is larger than the two side ones and often distinctly trifid. Ovate and membranous stipules are found at the base of the relatively long petioles.

Large and showy flowers are borne in 3-6-flowered umbel-like inflorescences with long and hairy peduncles. The flowers are cream-coloured with reddish stripes on the larger, upper two petals. Seven stamens bear anthers. Plants flower from September to January.

Distribution: *P. fragile* occurs from Worcester eastwards to the district of Peddie in the eastern Cape. It grows in rather dry and hot habitats and is often found on rocky slopes amongst karroid shrubs or bushes.

Remarks: This species is generally known as *P. tripartitum* due to the fact that Knuth (1912) disregarded the earliest specific epithet *fragile* of Andrews (1798).
P. trifidum of Jacquin (1797) is a later homonym and thus an illegitimate name.
P. fragile was introduced to England in 1792 by Lee and Kennedy. It can easily be propagated by cuttings.

×1

×2

×2

×1

Pelargonium fulgidum

(Celandine-leaved pelargonium; "Malva")

Fulgidus (Latin), shining or bright-coloured; refers to the brilliant-coloured flowers.

Shrub or scrambler, stem half-succulent. Leaves silvery and silky, stipules large and membranous.
Flowers scarlet or carmine with a prominent spur.

P. fulgidum (L.) L'Herit. in Ait. Hort. Kew. ed. 1,2: 422 (1789); Knuth in Pflanzenr. 4,129: 360 (1912).
Originally described by Linnaeus as *Geranium fulgidum* in Sp. Pl. ed. 1,2: 676 (1753).

Description: Small shrubby plant usually less than 1 m tall but when growing amongst other plants, supported branches may reach a height of more than 2 m. Stems half-succulent, becoming woody with age and covered with fairly persistent membranous stipules.
An attractive, silvery and silky foliage contributes to the beauty of this species. The oblong-cordate leaf blades are 3-10 cm long and 2-7 cm broad. They are pinnately deeply incised or 3-6-parted towards the base, with the lobes variously incised or

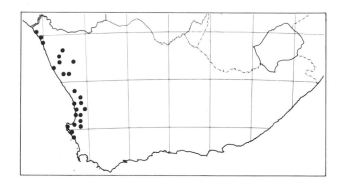

toothed. The petioles are 2-9 cm long and covered with long soft hairs (villous). Large asymmetrical, broadly ovate stipules are fused with the base of the petioles.
The peduncles are normally branched and relatively long. Scarlet or carmine flowers are borne above the foliage in 4-9-flowered umbel-like heads. A relatively long and prominent spur which expands to the calyx throat, contrasts strongly with the thinner pedicel. Seven fertile stamens are present. Plants flower from June to November.

Distribution: This species is confined to the western coastal districts and occurs from the vicinity of the Orange River to Ysterfontein. It grows on rocky, exposed wind-swept areas or sandhills near the coast.

Remarks: The history of this species dates back to the early days of colonization at the Cape. It was introduced to Holland early in the eighteenth century from where it was distributed to the garden of Pisa in Italy and the Sherard Garden near London. This species is also one of the parents of a number of beautiful hybrids.
The acidulous herbage is readily eaten by stock.

X1

X2

Pelargonium gibbosum

(Gouty pelargonium)

Gibbosus (Latin), unevenly swollen, refers to the swollen nodes of the stem.

Scrambler, stems with swollen nodes. Leaves glaucous, half-succulent. Flowers yellowish, night-scented.

P. gibbosum (L.) L'Herit. in Ait. Hort. Kew. ed. 1,2: 422 (1789); Knuth in Pflanzenr. 4,129: 361 (1912).
Originally described by Linnaeus as *Geranium gibbosum* in Sp. Pl. ed. 1,2: 677 (1753).

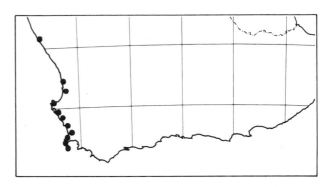

Description: The habit of this species is anything but typical of a pelargonium. It is often found scrambling amongst bushes, attaining a height of several metres. The stems are succulent when young, becoming woody with increasing age. It has been suggested that the swollen nodes store water but they could also act as means of support in that they lodge in the branches of the supporting plant.

The glaucous, half-succulent leaves are covered with a fine bloom similar to that found on a cabbage leaf. Their variable structure makes them difficult to describe. They are up to 13 cm long and 7,5 cm broad, usually 3- or 5-foliolate with one or two pairs of lateral leaflets and a terminal one. The lower pair of leaflets is usually petiolate and the upper ones sessile or even not entirely divided to the midrib. The margins of all the leaflets or segments are irregularly incised and toothed or lobed. Although the leaves are usually almost glabrous, the veins on the underside of the blade and margins of the leaflets may be hairy.

Inflorescences are umbel-like and resemble those of allied species such as *P. fulgidum* and *P. triste.* The peduncle with many short hairs and a few scattered long hairs, is 4-10 cm long and bears 6-14 flowers which may be dull ochre-yellow, barium yellow, but mostly greenish-yellow. The calyx segments are sharply reflexed and a spur, 2-4 cm long, originates at the posterior segment. Seven fertile stamens are present. Plants flower from November to April.

Distribution: *P. gibbosum* is frequent in sandy and rocky places near the sea. It is found in the False Bay area, in the Peninsula and along the West Coast as far north as Hondeklip Bay.

Remarks: A peculiar feature of this species is that it is night-scented. The flowers release a pleasant odour after sunset, probably to attract insects for pollination. It grows easily from cuttings and has been cultivated in Britain since 1712.
When growing in open veld the plants are often stunted, and also grazed.

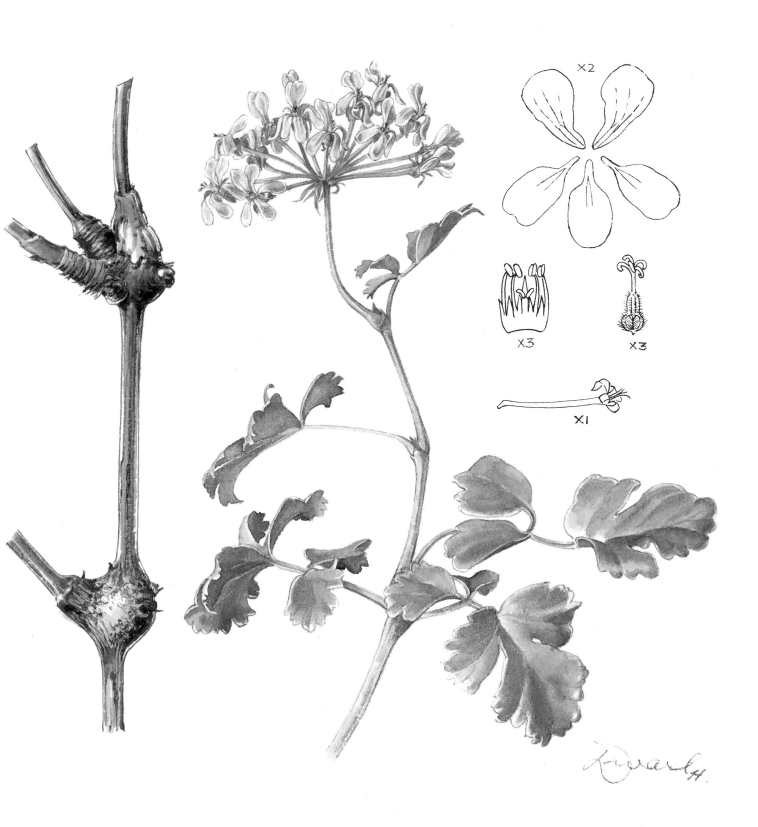

X2

X3

X3

X1

Pelargonium glaucum

(Spear-leaved pelargonium).

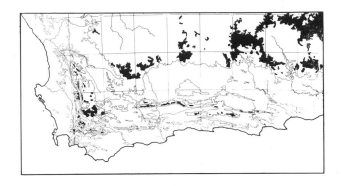

Glaucous (Latin), sea-green or covered with a bloom; applicable to the colour and surface of the leaves.

Halfshrub with smooth and glaucous shoots.
Leaves unifoliolate, glaucous, smooth and somewhat fleshy.
Flowers creamy white with a red blotch on upper petals.

P. glaucum (L.f.) L'Herit. in Ait. Hort. Kew. ed. 1,2: 429 (1789); Knuth in Pflanzenr. 4,129: 435 (1912).
Originally described by the son of Linnaeus as *Geranium glaucum* in L.f., Suppl.: 306 (1781).

Description: An erect and branched halfshrub up to 30 cm tall with the stems rather woody at the base. The smooth and glaucous side-branches are slender, brittle and conspicuously jointed.
The leaves are unifoliolate. This means that a leaf consists of a single leaflet which is distinctly articulated to the petiole. These entire leaflets are up to ca. 7,5 cm long and ca. 2,5 cm broad. They are glaucous, somewhat fleshy and lanceolate or spoon-shaped with a tapering point. The leaves are characteristically long-petioled with awl-shaped to linear stipules.
The cylindrical and glaucous 1-2-flowered peduncles are either axillary or opposite to the leaf. The colour of the petals varies from white to buff or even pale yellow with a red spot in the centre of the upper two petals which joins two red lines from the base. No blotch occurs on the smaller lower petals but they also have a red stripe from the base upwards. Seven fertile stamens are present. Plants bear flowers throughout the year.

Distribution: The species has a very limited distribution being confined to the Worcester district. This is an area with a relatively low annual rainfall. It grows in sandy and rocky places.

Remarks: This is a handsome pelargonium worthwhile growing in rockeries in drier parts of South Africa. Cuttings strike root readily. It was introduced into Britain by Kennedy and Lee in 1775.

P. tricuspidatum is a natural hybrid between *P. glaucum* and *P. scabrum.* These hybrids have been seen at Sanddrift near De Doorns.

Geranium glaucum which was described by Burman f. (1759) is not *P. glaucum.*

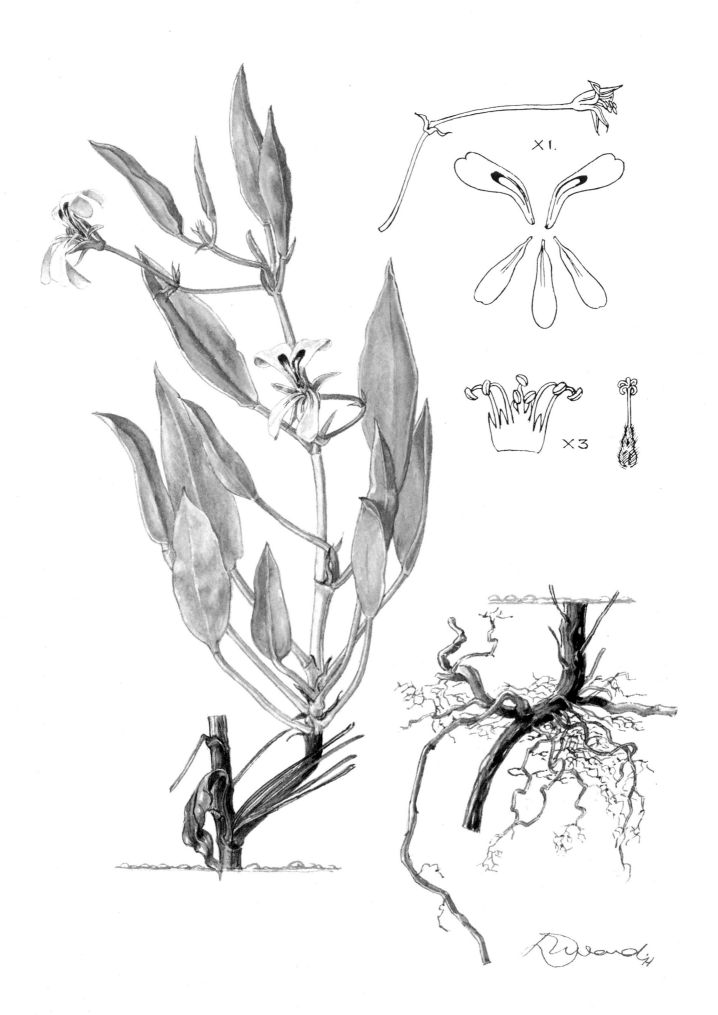

X1.

X3

Pelargonium grandiflorum

(Large-flowered pelargonium).

Grandi- (Latin), large; florum (Latin), flower.

Stems and leaves smooth and glaucous. Leaves 5-7-palmately-lobed, the upper ones being smaller and almost sessile. Flowers large and attractive.

P. grandiflorum (Andr.) Willd. in Sp. P1. 3,1: 674 (1800); Knuth in Pflanzenr. 4,129: 425 (1912).
Originally described as *Geranium grandiflorum* by Andrews in Bot. Rep.1: t. 12 (1799).

Description: An erect or straggling and herbaceous shrubby plant usually not more than 0,75 m tall. The soft stems are glaucous, smooth and shiny.
This is a species with an attractive glaucous and smooth foliage. The leaves are deeply 5-7-palmately-lobed, the lobes being divergent and acute with coarsely toothed margins. The lower leaves are usually larger (up

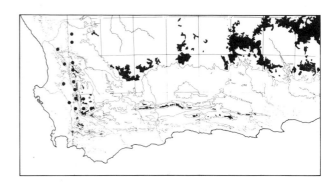

to 5 cm long and 8 cm broad) and longer petioled than the upper ones. All the leaves have a cordate base but only some leaves are zoned. The large and ovate stipules with ciliate margins are fairly conspicuous.
The peduncles are usually 2-3-flowered but up to 5-flowered inflorescences are also found. The colour of the relatively large and beautiful flowers varies from creamish-white to pink to purplish with darker streaks, and sometimes also blotches present on the upper two petals. The pollen of the seven fertile stamens is orange-coloured. It flowers from October to March.

Distribution: Confined to the South Western and Western Cape where it occurs from Nieuwoudtville to Tulbagh. It is usually found in mountainous habitats with an altitude of more than 300 m.

Remarks: Due to the larger leaves and flowers, *P. grandiflorum* resembles a giant form of *P. saniculaefolium*. No structural differences exist between the two taxa however, and it is possible that they are the same species.
This could be a fine garden plant in South Africa as it strikes readily from cuttings. It was introduced into England by Masson in 1794.

X1·5   X2

X1

X1

Pelargonium grossularioides

(Gooseberry-leaved pelargonium; "Rooirabas(sam)"; "Rooistingelhoutbas")

Grossularia (name of the Gooseberry); -oides (Greek), resemblance; the leaves resemble those of the Gooseberry.

Low spreading herb, stems long and reddish.
Leaves roundish to reniform, palmately lobed, aromatic.
Flowers small, usually purplish.

P. grossularioides (L.) L'Herit. in Ait. Hort. Kew. ed. 1,2: 420 (1789); Knuth in Pflanzenr. 4, 129; 410 (1912).
Originally described by Linnaeus as *Geranium grossularioides* in Sp. Pl. ed. 1,2: 679 (1753).

Description: A low spreading annual? herb branching from the base and attaining a height of ca. 20 cm. Individual stems may reach a length of up to 50 cm. They are characterized by their long, angular and furrowed reddish internodes.
The aromatic stems and leaves are almost glabrous to fairly pubescent with short hairs and glands. The leaves are roundish to reniform in outline, 3-5-palmately lobed and

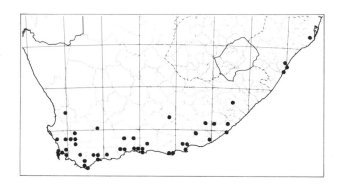

usually 1-4 cm long and 1-6 cm broad. Upper leaves are normally much smaller and more deeply incised. Triangular stipules are found at the base of the long and reddish petioles.

The relatively small and inconspicuous flowers are borne in 3-50-flowered umbel-like and compact inflorescences. The colour of the flowers varies from pink to beetroot purple but whitish flowers are occasionally found. Plants flower almost throughout the year. Seven fertile stamens are present.

Distribution: *P. grossularioides* occurs from Mozambique southwards along the coast to the South Western Cape. It is also recorded from Tristan da Cunha and as an alien from Kenya, India and California.
It is generally found in damp or shady places.

Remarks: This species shows great variability in the number of flowers per inflorescence as well as in the degree of incision and size of the leaves. Various varieties have been distinguished, though gradations in characters seem to be present.
Decoctions made from this species were formerly used by the Cape Malays to procure abortion, hasten confinement or to expel the placenta.

X3

X4

Pelargonium hirsutum var. melananthum

(Various-leaved pelargonium)

Hirsutus (Latin), hirsute; refers to the coarse and stiff hairs on the leaves and inflorescence; Melano- (Greek), black or very dark; refers to the colour of the flowers.

Geophyte, tuber ± globose.
Leaves polymorphous, hirsute.
Flowers dark purple, five fertile stamens.

P. hirsutum (Burm. f.) Soland. in Ait. Hort. Kew. ed. 1,2: 417 (1789); var. *melananthum* (Jacq.) Harv. in Fl. Cap. 1: 267 (1860); Knuth in Pflanzenr. 4,129: 340 (1912).
The species was originally described by Burman (filius) as *Geranium hirsutum* in Spec. Bot. Ger.: 50, t.68, f.2 (1759).

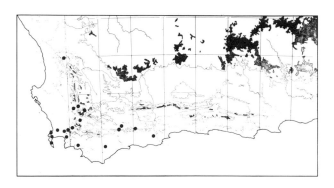

Description: An acaulescent geophyte with a height of 8-20 cm. The subterranean tuber is more or less globose and crowned with the dead bases of old petioles.
The structure of the radical leaves are extremely variable i.e. polymorphous (simple, pinnatifid, bi-pinnatifid or almost pinnatipartite). This variation may occur on a single plant. The leaves are covered with appressed coarse hairs (hirsute), clearly visible to the naked eye. The length of the leaf blades varies from 2-13 cm and the width from 1-5 cm. Petioles are usually longer than the blades. The lanceolate stipules are also hirsute.
Scapes, with the same indumentum as the leaves, are usually branched and they appear at a stage when at least some of the leaves start to wither. The relatively large and striking pseudo-umbels are 4-many-flowered. The petals are intense dark purple with a white base. Only five filaments bear anthers. The plants flower from September to December.

Distribution: This variety is confined to the South Western part of the Cape Province and occurs from Cape Town eastwards to Albertinia. It is often found in mountainous habitats.

Remarks: The varieties *melananthum* and *carneum* have similar leaves and are only distinguished on basis of floral characteristics. Var. *carneum* which occurs mainly in the Eastern Province, has white to rosy petals with red stripes.
Var. *melananthum* was introduced to Britain by Masson in 1788.

X1

X2   X3   X3

Pelargonium hirtum

(Hairy or fine-leaved pelargonium)

Hirtum (Latin), hairy, particularly with long distinct hairs; refers to the long hairs on the leaves.

Bushy shrublet with fleshy and armed stems.
Leaves resembling those of a carrot, villous.
Flowers relatively small, often magnolia purple.

P. hirtum (Burm. f.) Jacq. in Icon. Pl. Rar. 3: t.536 (1792); Knuth in Pflanzenr. 4,129: 381 (1912).
Originally described as **Geranium hirtum** by Burman (filius) in Spec. Bot. Ger.:48 (1759).

Synonym: *P. tenuifolium* L'Herit. in Ait. Hort. Kew. ed. 1,2: 421 (1789).

Description: A bushy shrublet attaining a height of 30 cm. The main stem is branched into erect or decumbent branches which are rather thick and fleshy. They have a greyish colour and are often armed with persisting bases of old petioles.
The finely divided compound leaves resemble those of a carrot. They are usually 1,5-3,5 cm long and 0,5-2 cm broad, covered

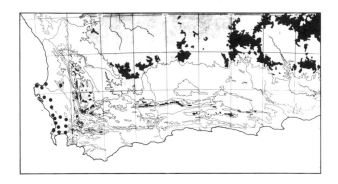

with long and soft hairs (villous), and are borne on long petioles. The leaflets occur in whorls on the rachis being subdivided into many narrow-linear segments. The lanceolate and sharp-pointed stipules are partly attached to the base of the petioles.
Flowering branches are conspicuously thinner than the ordinary branches; they are branched and often bear leaves. Flowers are borne in 3-8-flowered umbel-like inflorescences. The relatively small flowers have a bright pink to magnolia-purple colour. Darker spots occur at the base of the two obovate upper petals which are slightly larger than the lower three. Seven fertile stamens are present. This species flowers from July to November.

Distribution: *P. hirtum* has a limited distributional area, stretching from Velddrif to Stellenbosch. It is fairly common in the Cape Peninsula and usually grows in sandy soil but it is also found on rocky ledges.

Remarks: Plants with a similar habit as *P. hirtum* are found in the Pakhuis Pass near Clanwilliam. This is *P. oreophilum* which was described by Schlechter in 1900. Knuth (1912) considered *P. oreophilum* as to be conspecific with *P. hirtum*. The leaf characters, persistent petioles and flower of these two taxa differ so markedly that it seems justifiable to regard them as different species.
P. hirtum was introduced to England by Malcolm in 1768.

X1

X2

X2

Pelargonium inquinans

(Scarlet pelargonium, "Wilde malva").

The phrase "Folio digitis tacta inquinant colore ferrugineo", was given with the original description of the species. This means that the leaves become rusty or light brown on being touched. The common assumption that "inquinans" refers to the colour of the flowers, is therefore incorrect.

A softly woody shrub.
Leaves almost orbicular, cordate, velvety and glandular.
Flowers almost regular and with a long staminal column.

P. inquinans (L.) L'Herit. in Ait. Hort. Kew. ed. 1,2: 424 (1789); Knuth in Pflanzenr. 4,129: 441 (1912).
Originally described by Linnaeus as *Geranium inquinans* in Sp. Pl. ed. 1,2: 676 (1753).

Description: A moderately branched and softly woody shrub with a height of 1-2 m. The velvety and glandular branches are soft when young but harden externally with age.

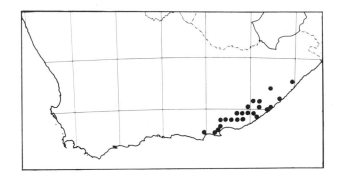

The almost orbicular leaves with a cordate base are normally 4-8 cm in diameter. Most leaves are shallowly 5-7-lobed with crenulate or finely toothed margins. The leaves have the same velvety texture and glands as the young stems. Only the young leaves have green, broadly-ovate stipules, the latter soon becoming membranous and wither with age.

The inflorescences are typically umbel-like and 5-30-flowered with the flower buds characteristically reflexed. The flowers are almost regular with the upper two petals only slightly smaller than the three lower ones. Intense scarlet flowers are most striking but pale pink or white flowers are also found. Seven fertile stamens and three staminodes are borne on a relatively long staminal column. Flowers are found almost throughout the year.

Distribution: This species is confined to the Eastern Cape and Transkei occurring from Patensie eastwards to Umtata. It is commonly found at the margin of coastal bush or further inland on the margin of succulent scrub usually in soil with shale predominating.

Remarks: *P. inquinans* is one of the parents of the garden hybrids known as the scarlet pelargoniums. It was cultivated in England by Bishop Compton as early as 1714.
Stems and leaves are pounded and used as a headache and cold remedy by tribespeople, also as a body deodorant.

X1

X3

Pelargonium lobatum

(Vine-leaved pelargonium; "Aandblom"; "Kaneelbol")

Lobatus (Latin), lobed; refers to the lobed leaves.

Geophyte, tuber large and irregular. Leaves large, usually 3-lobed, hairy. Flowers night-scented, petals purple-black with yellowish margin.

P. *lobatum* (Burm. f.) L'Herit. in Ait. Hort. Kew. ed. 1,2: 418 (1789); Knuth in Pflanzenr. 4, 129: 354 (1912).
Originally described by Burman (filius) as *Geranium lobatum* in Spec. Bot. Ger.: 44 (1759).

Description: A geophyte with a very large, irregularly shaped tuber clothed with a brown scaly bark. The leaves and inflorescences sprout from a very short stem. Both the stem and lower part of the petioles are often subterranean.
The leaves become extremely large, and their diameter may exceed 30 cm. The usually 3-lobed or 3-partite leaves are subdivided by a pattern of primary and secon-

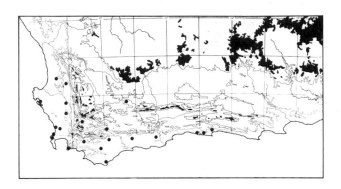

dary incisions to varying depths, in extreme cases giving rise to separate leaflets. Most leaves have a cordate base and an unequally toothed, ciliate margin. They are prominently veined, glandular and variably covered with soft hairs (villous to tomentose). Broadly ovate stipules are found at the base of the petioles.
The flowers are very fragrant at night and they are borne on branched scapes which may attain a height of up to 0,7 m. Typically star-like pseudo-umbels with 6-20 flowers each, develop. The petals are dark purple to almost black with a dull yellow-green margin and base. Six of the ten filaments bear anthers. The flowering period stretches from September to November.

Distribution: This interesting species is confined to the South Western and Southern Cape, occurring from Piketberg to the district of George.
It grows on sandy flats or against hillsides among other representatives of the Cape Fynbos.

Remarks: The scent emitted by the flowers towards the evening resembles cinnamon, hence the Afrikaans vernacular name "Kaneelbol".
This species can be propagated by the small tubers which develop on the original tuber.
It was cultivated in the Chelsea Garden (England) as early as 1739.

24

X2

X3

X1

X2

Pelargonium longifolium

(Bearded pelargonium)

Longi- (Latin), longer; -folium (Latin), leaf; refers to the length of the leaves.

Acaulescent geophyte.
Leaves extremely polymorphous, linear to bipinnate.
Flowers white, yellow or pink, five fertile stamens

P. longifolium (Burm.f.) Jacq. in Ic. Pl. Rar. 3: 11, t.518 (1792); Knuth in Pflanzenr. 4, 129: 322 (1912).
Originally described by Burman (filius) as *Geranium longifolium* in Spec. Bot. Ger.: 50 (1759).
Synonyms: *P. barbatum, P. ciliatum, P. depressum, P. longiflorum,* etc.

Description: An acaulescent geophyte with a subterranean, cylindrical or ovoid tuber. Inflorescences growing up from the tuber may attain a height of up to 25 cm. The roots are also tuberous.
The radical leaves are extremely polymorphous and vary even on the same plant. Leaf blades are linear, lanceolate or ovate with an entire margin. They can also be

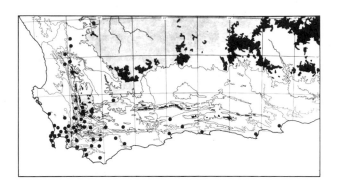

shallowly to deeply incised, or even bipinnately divided into linear segments. The leaves are glabrous to pubescent, and the length of the blades varies from 1-12 cm and the width from 0,3-7 cm. Narrowly lanceolate to filiform stipules are present.
Flowering stems are hairy and branched to bear two or more 3-5-flowered pseudo-umbels. The colour of the petals varies from white to yellow to pink with dark purple markings. The two larger upper petals are in some cases blotched with dark purple. Of the five filaments which bear anthers, one is distinctly shorter than the others. Flowers are found throughout the year, being most abundant during the dry hot summer months of November, December and January.

Distribution: *P. longifolium* is more or less restricted to the winter rainfall area of the Cape Province, and it has roughly the same distribution as Fynbos or Macchia. It occurs from Calvinia to Port Elizabeth, being very abundant in the South Western Cape.
It usually grows in rocky or sandy soil.

Remarks: Several species described in the past, are now considered as synonyms of *P. longifolium.* When taking into consideration the variability of leaf forms and floral characteristics in this species, it can be understood why several species had been distinguished. Only intensive field work will finalize the merits of these synonyms.
Drawing A illustrates the synonym *P. barbatum.*

Pelargonium luridum

Luridus (Latin), smoky or drab yellow; refers to the colour of the flowers.

Acaulescent geophyte, tuber woody.
Leaves extremely polymorphous.
Scape long, flowers large and of variable colour.

P. luridum (Andr.) Sweet in Colv. Cat. ed. 2: 2 (1822).
Originally described by Andrews as *Geranium luridum* in Geran. 2: t. 34 (1805).

Synonym: *P. aconitiphyllum* (Eckl. & Zeyh.) Steud. in Nom. Bot. 2: 283 (1841); Knuth in Pflanzenr. 4, 129: 361 (1912).
Müller (1963) listed several other synonyms.

Description: An acaulescent geophyte with a woody, subterranean tuber. Conspicuous inflorescences growing up from the rootstock, may attain a height of 1 m. The vegetative parts of the plant as well as the inflorescence are variably covered with long hairs and sessile glands.
This species is remarkable for its variability of leaf-shape on a single plant. The first leaves produced each season are shallowly

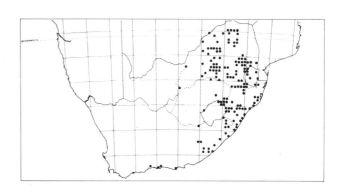

lobed to pinnatisect or bipinnatisect. Successive leaves become larger and progressively more dissected with narrower segments which can ultimately be filiform (in the Transvaal). All the leaves are radical with the leaf blades up to 27 cm and petioles up to 30 cm long. The stipules are linear to narrowly triangular.
The long scape ends in a large pseudo-umbel with 5-60 flowers. The colour of the relatively large flowers varies from white to pink, yellow to greenish-yellow or may even show variegations on a single petal. Red flowers are recorded from Mozambique and Angola. Seven filaments are fertile. The flowering period extends from September to April.

Distribution: This interesting pelargonium has a very wide distribution. In South Africa it occurs in the Eastern Cape, Natal, Orange Free State and Transvaal. It is also recorded from Lesotho, Swaziland, Angola, Rhodesia, Malawi, Mozambique, Zambia, Zaire and Tanzania.
P. luridum is usually found in damp habitats in grassveld.

Remarks: The Zulu people use an infusion of the root for diarrhoea. They also mix dried powdered root in porridge or other food in the treatment of dysentery. During courtship, the young men rub a mixture of this root powder and fat of hippopotamus or python on their faces to charm the opposite sex.

X1

X3

X1.5

Pelargonium myrrhifolium
var. betonicum

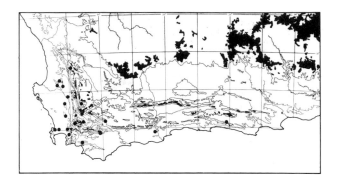

Myrrh, vernacular name of members of the Burseraceae; -folium (Latin), leaf. *Betonica,* a genus of the Labiatae; the leaves resemble those of *Betonica.*

Small, erect halfshrub.
Leaves pinnately incised, hairy.
Flowers white or pink, five fertile stamens.

P. myrrhifolium (L.) L'Herit. var. *betonicum* (Burm. f.) Harv. in Fl. Cap. 1: 286 (1860); Knuth in Pflanzenr. 4, 129: 395 (1912).
The species was originally described by Linnaeus as *Geranium myrrhifolium* in Sp. Pl. ed. 1,2: 677 (1753), and the variety by Burman (filius) as *Geranium betonicum* in Spec. Bot. Ger.: 32 (1759).

Description: A small, erect halfshrub attaining a height of ca. 30 cm. The branches are covered with both glandular hairs and coarse non-glandular hairs (hirsute). The plant has a well developed taproot system.

The leaves are pinnately incised (pinnatifid or bipinnatifid) with an elongate cordate outline, and they have a similar indumentum as the stems. Broadly ovate stipules with a ciliate margin, are found at the base of the petioles.
The inflorescences are 2-6-flowered and umbel-like. The colour of the four or five petals varies from white to pink with beet-root-red veins in the upper two petals. Five filaments bear anthers. It flowers sporadically throughout the year reaching a definite peak in Spring.

Distribution: This variety occurs from Krom River in the Cedar Mountains to the Lang Kloof. It is very well represented in the South Western Cape.
It usually grows in clayey soil.

Remarks: Var. *betonicum* is one of the nine varieties of *P. myrrhifolium* described by Harvey in 1860.
Unfortunately the name of this variety must be changed to var. *myrrhifolium,* because it was this taxon which was originally described by Linnaeus as *P. myrrhifolium.*
It was introduced to Britain by Philip Miller in 1731.

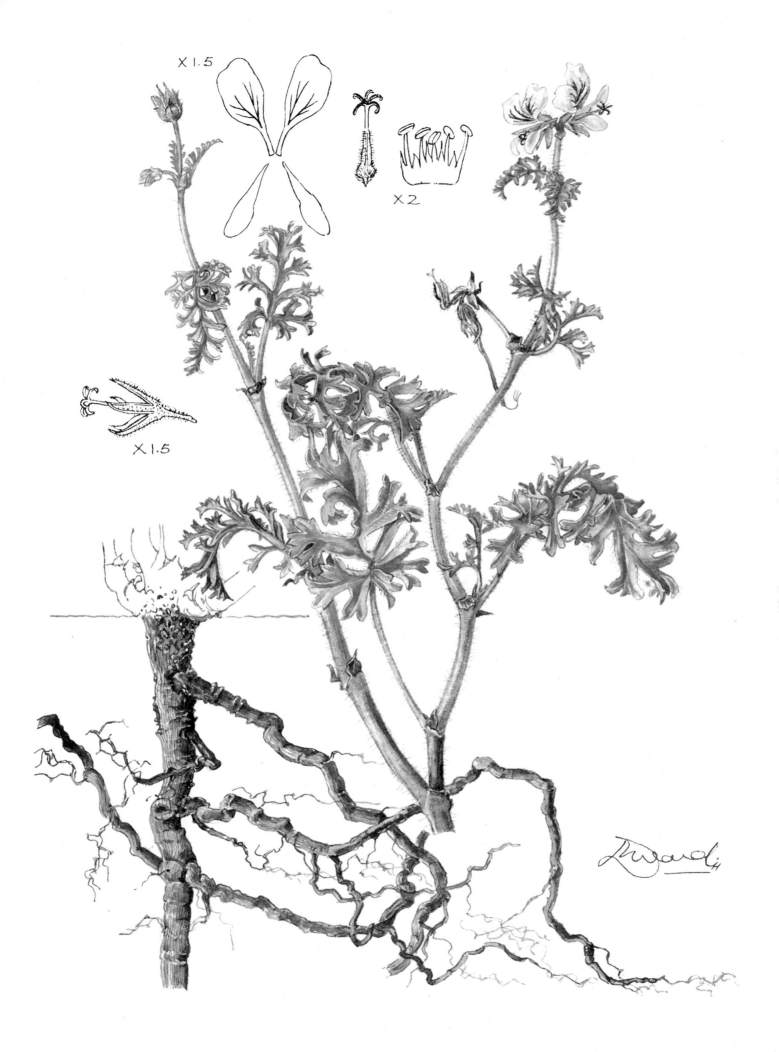

X 1.5

X 2

X 1.5

Pelargonium myrrhifolium var. fruticosum

(Myrrh-leaved pelargonium).

Myrrh, vernacular name of members of the Burseraceae; -folium (Latin), leaf. Fruticosus (Latin), shrubby.

Small halfshrub, branches procumbent.
Leaves bipinnately incised, glabrous or nearly so.
Flowers white or pale pink, papilionaceous, seven fertile stamens.

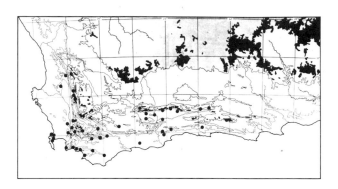

P. myrrhifolium (L.) L'Herit. var. *fruticosum* (Cav.) Harv. in Fl. Cap. 1: 286 (1860); Knuth in Pflanzenr. 4, 129: 395 (1912).
The variety was originally described by Cavanilles as *Geranium fruticosum* in Cav. Diss. 4: 263, t. 122, f.2 (1787).

Description: An attractive and shrubby plant attaining a height of 30-40 cm. The branches are procumbent and covered with both short stiff hairs and longer softer hairs (villous). The leaves have an elongate cordate shape and their margin is incised to the midrib in a bipinnate way (bipinnatisect). It seems, however, as if the margin of the older leaves of the Table Mountain form is less incised. The leaf segments tend to curve upwards and they are often tinged red at the tips. The leaves are either glabrous or covered with short hairs.
The inflorescences are 1-4-flowered and umbel-like. Plants occurring in the drier parts of the Southern Cape have only 1-2 flowers per inflorescence. The colour of petals varies from white to pale pink with beetroot-red markings on the upper two petals. These two petals are much larger than the lower two (or three). Seven fertile stamens are present. It flowers from August to January with a definite peak during Spring.

Distribution: This variety is well represented in the Western Cape, but is found as far north as Wupperthal and as far east as Willowmore.
It usually grows in sandy soil.

Remarks: This variety makes a very attractive subject for a rockery or border and deserves to be more widely known in South African gardens.

X1

X2

Pelargonium oblongatum

Oblongus (Latin), oblong; apparently refers to the oblong tuber.

Geophyte, oblong tuber partly epiterranean.
Leaves usually cordate ovate.
Flowers large, pale yellow.

P. oblongatum E. Mey. ex Harv. in Fl. Cap. 1: 263 (1860) — original description; Knuth in Pflanzenr. 4, 129: 329 (1912).

Description: A peculiar, acaulescent geophyte with an oblong vertical tuber which is partially exposed above ground level. The tuber can attain a length of 15 cm, it often has a flaking bark and is crowned with the remains of old stipules and petioles.

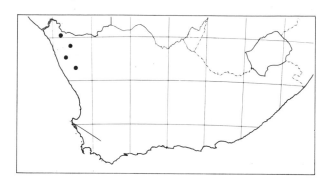

The shape of the leaf blades varies from cordate ovate to orbicular or even obovate. They are 3-10 cm long and 2-8 cm broad, rather fleshy, glandular and variously covered with coarse hairs. The crenate or lobulate margin is distinctly ciliate. The hairy petioles are usually relatively short, but the lanceolate stipules can reach a length of 2,5 cm.
This species has a tall and attractive inflorescence with branched and hairy scapes. Large, pale yellow flowers are borne in pseudo-umbels. The two upper petals with red-purple veins are larger and more distinctly obovate than the three lower petals. The five long filaments which bear anthers are bent towards the upper petals. It flowers from September to November.

Distribution: This interesting pelargonium is restricted to the Richtersveld and the northern part of Namaqualand. It has been collected only from Stinkfontein to the Kamies Mountains (near Kamieskroon). The climate of this area is hot and arid, with a low annual rainfall.

Remarks: E. Meyer named this species from a collection made by Drège, but apparently Burchell was the first to discover it earlier in the year 1814.
It was described for the first time by Harvey in 1860.

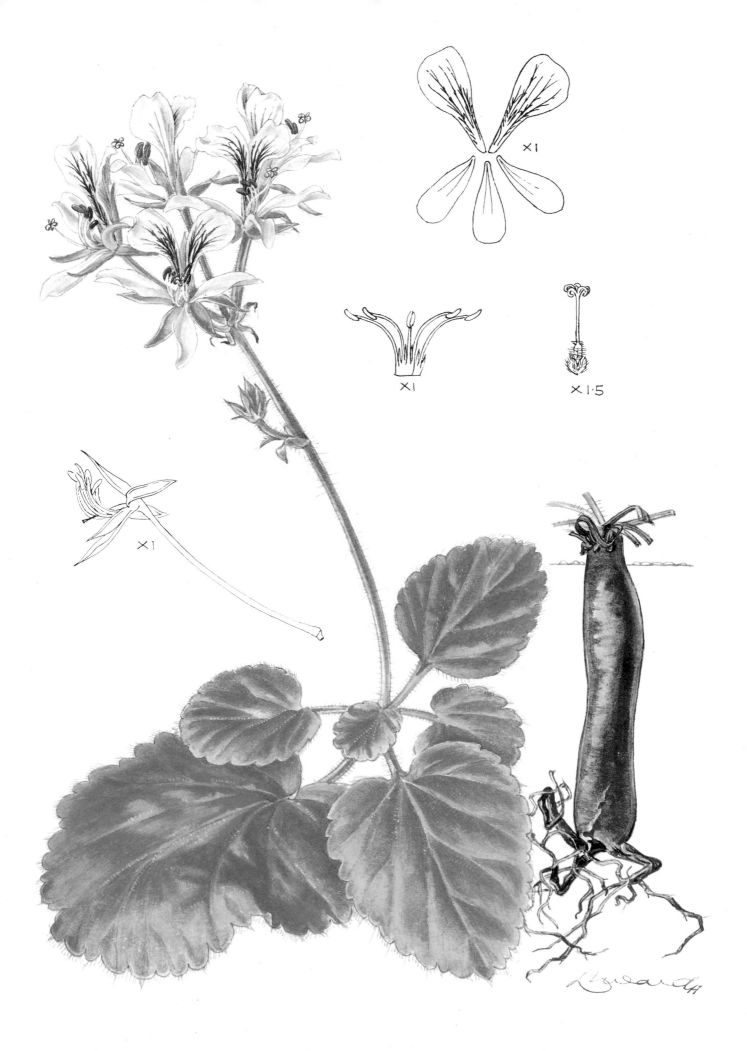

×1

×1

×1.5

×1

Pelargonium odoratissimum

(Sweet-scented pelargonium)

Odoratus (Latin), fragrant; refers to the sweet smell of the leaves.

Small perennial, prostrate shrublet.
Leaves roundish- or ovate-cordate with an apple-mint scent.
Flowers small and usually white.

P. odoratissimum (L.) L'Herit. in Ait. Hort. Kew. ed. 1,2: 419 (1789); Knuth in Pflanzenr. 4, 129: 452 (1912).
Originally described as *Geranium odoratissimum* by Linnaeus in Sp. Pl. ed. 1,2: 679 (1753).

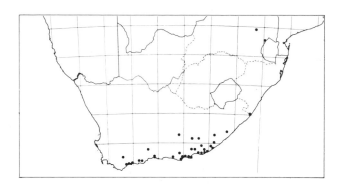

Description: A perennial and prostrate shrublet with a short, thick main stem and sprawling, herbaceous flowering branches which may reach a length of about 60 cm. The height of the plant rarely exceeds 30 cm. The main stem is rough and scaly due to persisting bases of old stipules. The roots are slightly tuberous.
The roundish- or ovate-cordate leaves with obtusely crenulate margins are usually 3-4 cm in diameter. Leaves on the main stems are normally much larger (up to 12 cm in diameter) than those on the elongated flowering stems. The leaves have an apple-green colour and they are covered with fine, short hairs (pubescent), making them soft to the touch. A strong, sweet apple-mint scent is perceptible when the leaves are bruised. Small deltoid stipules are found at the bases of the very long petioles.
The flowers borne in 3-10-flowered umbel-like inflorescences, are relatively small, their petals being slightly longer than the sepals. In general the flowers are white with crimson markings on the upper two petals. Pale pink flowers are occasionally found. Seven fertile stamens are present. Plants flower almost throughout the year with the exception of the midsummer months.

Distribution: This species is common in the Eastern and Southern Cape, but it is also recorded from the Lowveld of Transvaal and Natal. It occurs as undergrowth in forests or in shady places protected by bushes or rocky ledges.

Remarks: A related species, *P. mollicomum*, was described by Fourcade in 1932. He mentioned that it differs from *P. odoratissimum* in the much coarser pubescence of the leaves, the ciliate membranous stipules, the linear calyx lobes and the much larger flowers. Further investigation could prove that these two species are in fact conspecific.
P. odoratissimum was introduced into Britain in 1724 where it was cultivated in the renowned Chelsea Garden.

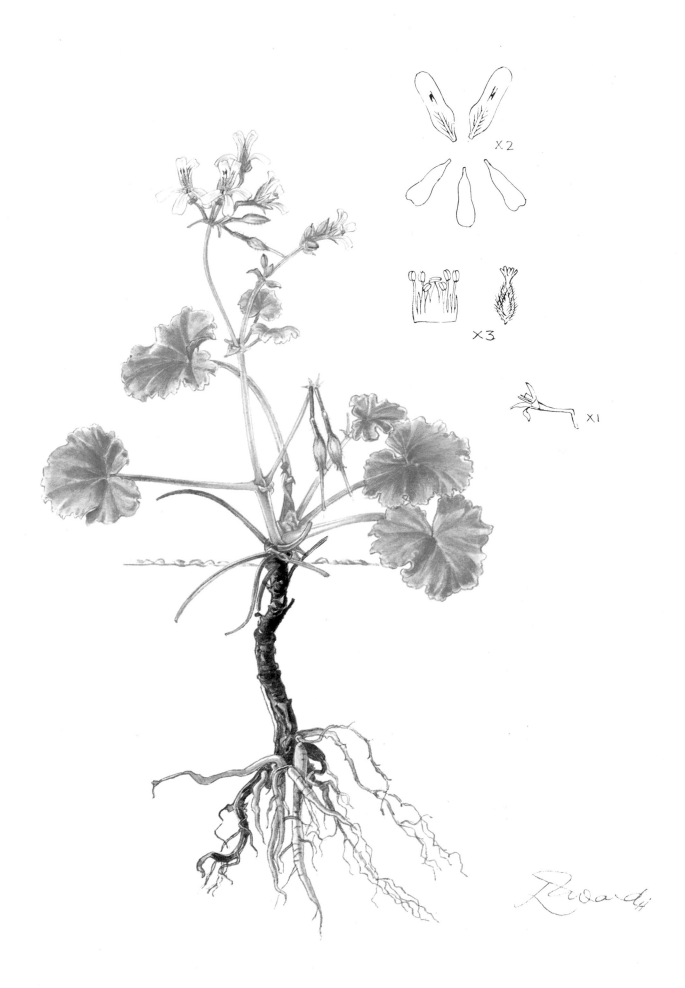

X.2

X3

X1

Pelargonium ovale

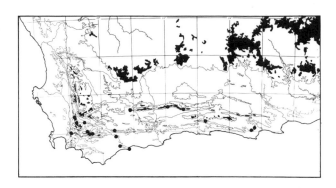

Ovalis (Latin), oval; refers to the shape of the leaves.

Low growing halfshrub.
Leaves usually ovate, grey-green, hairy, stipules persistent.
Flowers pale to dark pink, five fertile stamens, two staminodes recurved.

P. ovale (Burm. f.) L'Herit. in Ait. Hort. Kew. ed. 1,2: 429 (1789); Knuth in Pflanzenr. 4,129: 416 (1912).
Originally described by Burman (filius) as *Geranium ovale* in Prod. Fl. Cap. 1: 19 (1768).

Description: A perennial low growing halfshrub reaching a height of ca. 30 cm. The stems are branched and covered with persistent stipules. Plants often exhibit a cushion habit.
The petioles of the grey-green leaves are usually twice the length of the leaf blades. Leaf blades are typically oval though gradations to a semi-round shape occur. They measure 2,5-3,5 cm in length and 1,5-2,5 cm in width, and are covered with a villous to tomentose indumentum. The hairs are prominent on the raised veins of the under surface, and the margin of the leaves is serrulate-crenulate. Deltoid-acuminate, persisting stipules form a sheath around the base of the petioles.
Flowering stems are branched and very floriferous. Each of the villous peduncles bears 3-5 flowers. The glandular calyx elongates with age and turns a russet-brown. The colour of the flowers varies from pale to dark pink with darker markings on the two upper petals. Five fertile stamens with orange-coloured anthers are present. Two of the five staminodes are typically recurved. It flowers throughout the year with a peak period lasting from spring to summer.

Distribution: The distribution range of this species extends from the South Western Cape to the Swartberge and eastwards to Humansdorp. It is well represented in the mountain and coastal Fynbos, where it grows in well-drained sandy soil.

Remarks: The name of the section Campylia apparently pertains to the two, typically recurved staminodes of this species (Kampylo (Greek), bent or curved).
Four varieties of P. ovale are usually distinguished. The variety described and illustrated in this work is known as var. *blattarium.*
Its cultivation in England dates from the mid eighteenth century.

X1.

X2

Pelargonium papilionaceum

(Butterfly pelargonium; "Rambossie")

Papilionaceus (Latin), butterfly-like; refers to the flower with two large upper petals, resembling a butterfly.

Erect and strongly aromatic shrub.
Leaves cordate, shallowly lobed.
Flowers light pink to carmine, papilionaceus.

P. papilionaceum (L.) L'Herit. in Ait. Hort. Kew. ed. 1,2: 423 (1789); Knuth in Pflanzenr. 4,129: 465 (1912).
Originally described by Linnaeus as *Geranium papilionaceum* in Sp. Pl. ed. 1,2: 676 (1753).

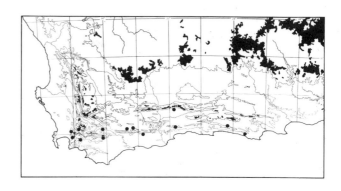

Description: An erect, much-branched and strongly aromatic shrub attaining a height of more than 2 m, although usually ca. 1 m tall. The main stem is woody at the base, while the side-branches are herbaceous and covered with long soft hairs (villous).
The cordate or roundly cordate leaves are usually ca. 7 cm long and ca. 10 cm broad, the upper ones being entire and the lower ones shallowly 3-to 5-lobed. The leaves are glandular and partly covered with long hairs. They are more strongly scented and usually more conspicuously veined on the upper side than the leaves of *P. vitifolium*. The margins are finely toothed, serrulated or almost entire. Broadly ovate stipules, sometimes with teeth, are found at the base of the long, villous petioles.
The branched peduncles form many umbel-like heads with 5-12 flowers each. Flowers are borne on rather long and villous pedicels. They are most attractive with two large reflexed upper petals and three very narrow lower ones. The colour of the petals varies from light pink to carmine. A dark purple blotch on the upper petals, contrasts well with an adjacent white blotch. Seven fertile stamens are present. Plants flower from Spring to January.

Distribution: This species is found in the South Western, Southern and Eastern Cape, and occurs from the Stellenbosch district eastwards to Grahamstown. It grows in kloofs and at the margins of indigenous forests in half-shady places, usually near streams.

Remarks: *P. papilionaceum* is a fascinating species with pretty flowers, worth-while growing in shady places of a garden. It grows easily from cuttings. It has been cultivated in Britain since 1724.
The leaves have an unpleasant odour, described by some as that of a he-goat — hence the vernacular name "Rambossie". It has been reported that the leaves were used as a tobacco substitute, and it is possible that they were smoked for medicinal purpose.

×1·5

×2

Pelargonium peltatum

(Ivy-leaved pelargonium; "Kolsuring").

Peltatus (Latin), peltate; refers to the shield-like leaves, the petiole being attached to the centre of the blade.

Creeper with smooth and slender stems. Leaves peltate, fleshy, 5-angled or lobed and sometimes zoned.
Flowers long-spurred and with five petals.

P. peltatum (L.) L'Herit. in Ait. Hort. Kew. ed. 1,2: 427 (1789); Knuth in Pflanzenr. 4, 129: 422 (1912).
Originally described by Linnaeus as *Geranium peltatum* in Sp. Pl. ed. 1,2: 678 (1753).

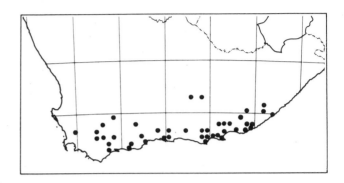

Description: A climbing, herbaceous perennial trailing into and through other bushes, attaining a height of up to 2 m or more. The long straggling shoots are smooth and slender with a diameter of 0,3-1 cm.
This interesting pelargonium has the same peltate or shield-like leaves as the nasturtium. The leaves are succulent and bluntly 5-angled or lobed with entire margins. They are 2-7 cm in diameter and usually more or less glabrous although hairy leaves are common in plants from the Eastern Cape.

The petioles are usually slightly shorter than the leaf blades and broadly ovate stipules are found at their base. Zonation of the leaves may occur but this is not a constant characteristic.
The flowers are borne in 2-9-flowered umbel-like inflorescences which are long pedunculated. The colour of the flowers varies from mauve or pinkish mauve to pale pink or even whitish, while only the two upper petals show darker markings. Two of the seven fertile stamens are relatively short. It flowers almost throughout the year but masses of flowers are found during Spring.

Distribution: Occurs in the Cape Province from Wellington to East London; commonly growing in sheltered places with coastal or succulent bush.

Remarks: The leaves are rich in an acidulous sap, said to act as an effective astringent and antiseptic for sore throats. Juice from petals yield a blue indigo and it is recorded that Burchell used this in painting.
P. peltatum which is the ancestor of many garden hybrids was introduced into Holland by Willem Adriaan van der Stel in 1700, and into Britain by Masson in 1774.
Three varieties are recognized but they all grow intermixed. The closely allied *P. lateripes* which grows in the Transkei, Natal and Eastern Transvaal differs only from *P. peltatum* by having cordate leaves, that is, the petioles are inserted at the margin of the leaf.

×1

×1

×2

Pelargonium pinnatum

(Pinnated pelargonium).

Pinnatus (Latin), pinnate; refers to the leaves with leaflets arranged along a common axis, the rachis.

Acaulescent plant with a subterranean tuber.
Leaves radical and pinnate, leaflets mostly oval and hirsute.
Inflorescence branched, flowers with five fertile stamens.

P. *pinnatum* (L.) L'Herit. in Ait. Hort. Kew. ed. 1,2:417 (1789); Knuth in Pflanzenr. 4, 129: 347 (1912). Originally described by Linnaeus as *Geranium pinnatum* in Sp. Pl. ed. 1,2: 677 (1753).

Description: An acaulescent (apparently stemless) plant with a subterranean tuber. Several heads can develop with age from the original globose or cylindrical tuber. The roots are also tuberous.
Radical and pinnate leaves with a variable number of leaflets are borne on long and villous petioles. The leaves vary in length from 4-25 cm and the number of leaflets per leaf from 5-80. The leaflets are mostly almost sessile and occur alternately or almost opposite on another on the rachis. They are

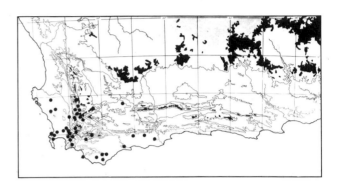

roundish, oval or elliptic, usually ca. 8 mm long and ca. 5 mm broad, and covered with long coarse hairs (hirsute) easily visible to the naked eye.
Long and villous inflorescences appear at a stage when the leaves are withering. The peduncles are usually branched to form 2-17-flowered umbel-like heads. The flowers vary in shape and colour. They are white, pale yellow, pale pink, salmon or flesh-coloured with a fine veining or a darker spot on the upper petals. Only five fertile stamens are present. The plants flower from September to April.

Distribution: The species is restricted to the south western part of the Cape Province, occurring from Hopefield south-eastwards to Albertinia. It is very common on the mountains near Ceres, Worcester, Stellenbosch and Cape Town.

Remarks: P. *pinnatum* is a very variable species and it seems likely to be conspecific with the so-called P. *astragalifolium* (illegitimate name!), being characterized by narrow and undulated petals. Drawings A and B are typical examples of what is usually considered as P. *astragalifolium* and P. *pinnatum* respectively.
This is an attractive species but requires a particular kind of management to grow. It should be planted in small, well-drained pots and must be kept quite dry during the dormant period.
It was introduced into Britain by Masson in 1788.

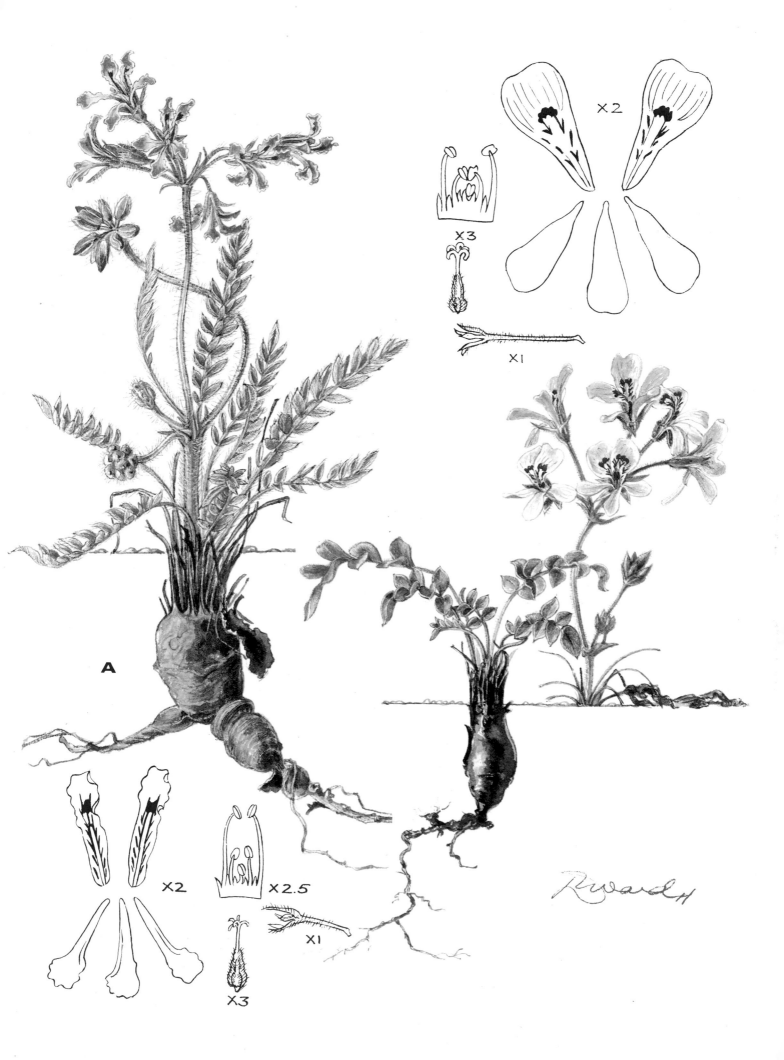

A

X2

X3

X1

X2

X2.5

X3

X1

Pelargonium praemorsum

(Five-fingered pelargonium; Quinate-leaved pelargonium)

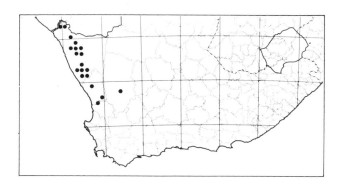

Prae- (Latin), before; morsus (Latin), a bite; refers to the toothed margin of the leaf segments which appears to be bitten.

Shrubby, stems distinctly jointed.
Leaves palmately 5-parted, apex of segments 3-toothed.
Flowers large, cream-coloured.

P. praemorsum (Andr.) Dietrich in Lex. Gärt. Bot. 7: 48 (1807).
Originally described by Andrews as *Geranium praemorsum* in Andr. Bot. Rep. 3: t. 150 (1801).

Synonym: *P. quinatum* Sims in Bot. Mag. 15: t. 547 (1802).

Description: A shrubby plant which may attain a height of 1 m, though usually much smaller. The flexuose and distinctly jointed stems have a shiny brown bark, and they are sparsely covered with very short hairs.
The leaves are kidney-shaped and palmately 5-lobed or more often deeply divided into five segments. Leaf blades as a whole are 0,3-2 cm long and 0,5-3 cm broad. The segments are wedge-shaped, generally with a three-toothed apex, and finely pubescent. The outer two segments are normally larger than the other three. Rigid, triangular stipules as well as the lower part of the petioles, are usually persistent.
The erect peduncles with membranous and conspicuous bracts at their base, are normally 1-2-flowered. The large attractive flowers are cream-coloured. Reddish-purple streaks decorate the two upper petals, which are much larger than the three (or two) lower ones. Flowers can be found from August to April. Seven filaments bear anthers.

Distribution: This is another pelargonium which is confined to Namaqualand, occurring from the Richtersveld southwards to the Van Rhynsdorp area. It often grows in the shelter of rocks or other plants.

Remarks: This species is more commonly known as *P. quinatum*, but the specific name *praemorsum* has priority over the name *quinatum*.
It was first raised from seed in England by Quarrell in the Chelsea Garden of Colvill. Cuttings root readily.

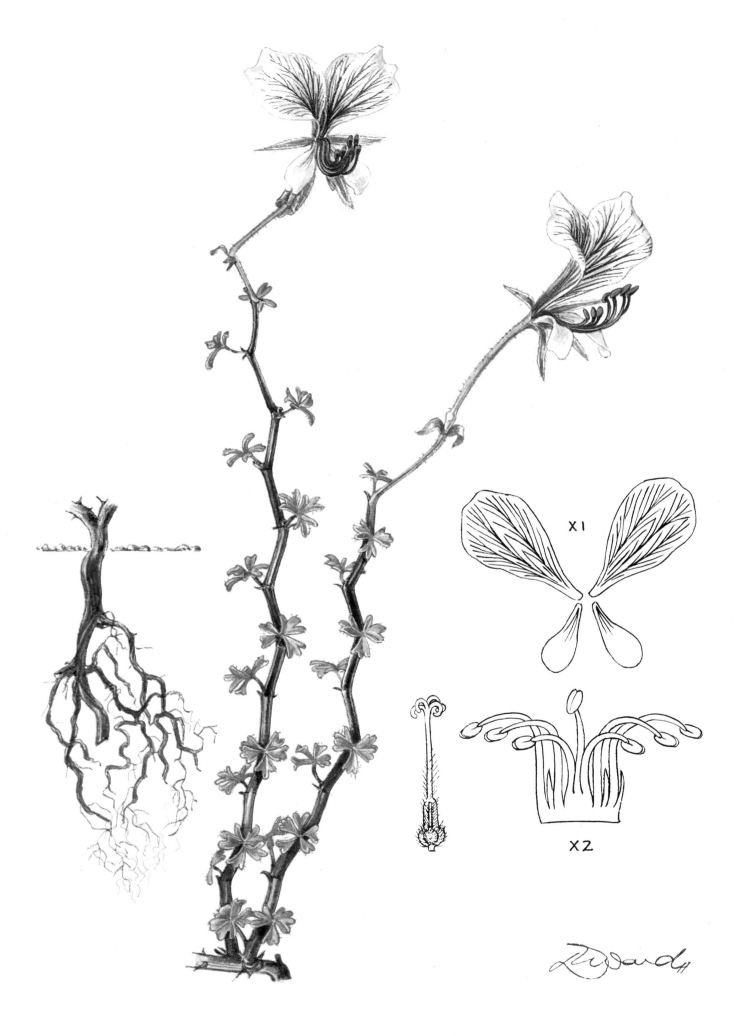

X1

X2

Pelargonium pulchellum

(Nonesuch pelargonium)

Pulchellus (Latin), beautiful and little.

Small halfshrub, stem short and fleshy. Leaves pinnately incised, hairy, stipules large. Petals white with red blotches and stripes.

P. pulchellum Sims in Bot. Mag. 15: t.524 (1801) — original description; Knuth in Pflanzenr. 4, 129: 379 (1912).

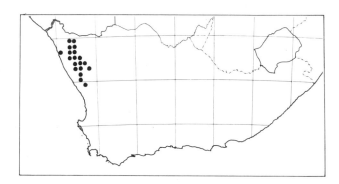

Description: A very small and attractive halfshrub with a conspicuous inflorescence which attains a height of 50 cm. The succulent stem is short, semi-procumbent and sparsely-branched.
The leaves are oblong ovate and irregularly pinnately incised. Leaf blades are 3-15 cm long and 1,5-6 cm broad, glandular and also covered with coarse appressed hairs. The petioles are relatively short and decurrent along the stem. Large, lanceolate stipules are fused to the lower part of the petioles. This lower part of the petioles with the fused stipules, is often persistent.
The branched scapes are densely covered with long hairs and they bear 6-25-flowered pseudo-umbels. The pure white flowers, marked with a red blotch or stripes, are singularly striking. There are instances where the upper two petals are only faintly veined in red, without any blotches. Seven stamens are fertile. It flowers from July to October.

Distribution: This is another beautiful and interesting pelargonium from Namaqualand, occurring from Klipfontein, north of Steinkopf, southwards to Nuwerus. It grows on dry granite outcrops in full sunlight.

Remarks: Vorster (1973) gives a lengthy discussion on the nomenclatural history of this species.
A plant was sent to Kew by Masson in 1795. It is easily propagated by cuttings, provided that it is watered extremely sparsely.

X2·5

X2

X1

Pelargonium pulverulentum

(Powdered-leaved pelargonium)

Pulverulentus (Latin), powdered; refers to the powdery pubescence of the leaves.

Geophyte, elongate tuber with a cracked bark.
Leaves glaucous, half-succulent, margin ciliate.
Petals yellow, often with dark blotches.

P. pulverulentum Colv. ex Sweet in Geraniaceae 3: 218 (1824) — original description; Knuth in Pflanzenr. 4,129: 355 (1912).

Synonym: *P. hollandii* Leighton in S. Afr. Gdng. Country Life 22: 232 (1932).

Description: This geophyte has a large, elongate, subterranean tuber with a brown, cracked bark. A short stem, bearing a few leaves and a conspicuous inflorescence, develops seasonally from this tuber.
The glaucous, half-succulent leaves are covered with a powdery pubescence. The leaf blades 4-10 cm long and 3-9 cm broad, vary considerably in shape. They are cordate ovate to oblong-ovate, shallowly lobed

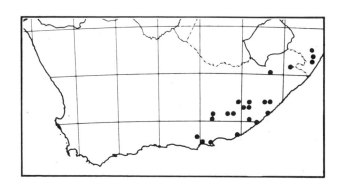

to deeply indented and variably covered with coarse hairs. The long soft hairs of the ciliate margin are readily visible. Triangularly shaped stipules are found at the base of the long petioles.
The branched and hairy inflorescences may attain a height of up to 0,5 m. Three to fourteen flowers are borne in a typical star-like pseudo-umbel. The two upper petals are usually reflexed, thus appearing to be smaller than the three lower ones. The five pale yellow petals can be either without any blotches, or blotched with dark purple to brown. Some flowers have the three lower petals mainly dark-coloured, with a mere edging of yellow . Six filaments bear anthers. It flowers from September to February.

Distribution: This species occurs in the Eastern Cape, Transkei and Natal. Growing in sandy soil, it is often a minor component of the grassveld communities of these areas.

Remarks: Superstitious Xhosa warriors believed, during the wars of 1850-1853, that by pointing the roots of these plants at the guns of their enemy, they would cause either the dampening of the gunpowder or ward off the bullets. They made use of the roots as an anti-diarrhoetic or antidysenteric medicine, while the leaves were regarded as a haemorrhoid curative.
It has been recorded that Colvill (England) received several plants of this species from the Cape in 1822.

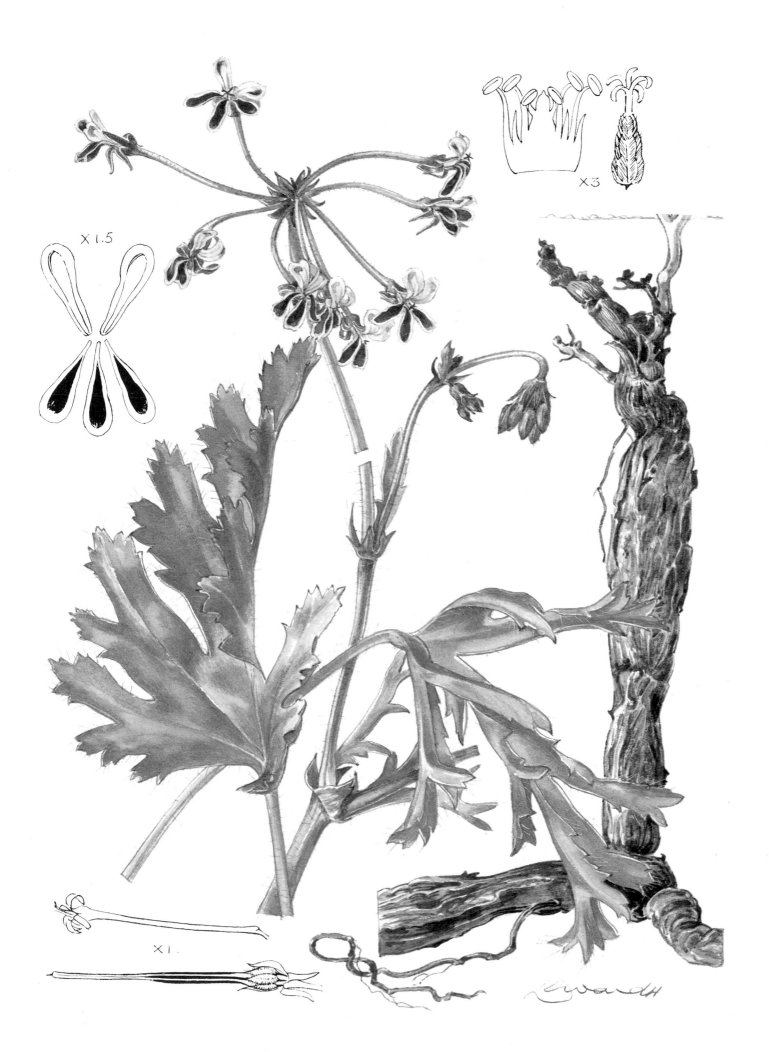

X3

X1.5

X1.

Pelargonium radens

(Rasp-leaved or multifid-leaved pelargonium).

Radens (Latin), scraping; Radula (Latin), rasp or file; refers to the rasp-like leaves.

An erect and densely-branched shrub. Leaves scabrid and deeply divided into narrow, obtusely lobulate segments with margins rolled under.

P. radens H.E. Moore in Baileya 3,1: 22 (1955).

Synonym: *P. radula* (Cav.) L'Herit. in Ait. Hort. Kew. ed. 1,2: 423 (1789); Knuth in Pflanzenr. 4,129: 477 (1912).

Description: An erect, densely-branched shrub usually less than 1 m tall, with a stem becoming woody toward the base with age. The side-branches remain herbaceous and slender and are covered with stiff bristles making them rough to the touch.

The much and deeply divided (palmately-bipinnatifid) leaves are triangular in outline, 3-5 cm long and 3-6,5 cm broad. The scabrid leaves have a pungent odour when bruised. The narrow segments with margins

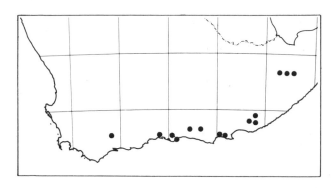

characteristically rolled under and rounded apex, are usually obtusely lobulate. Narrowly ovate or ovate stipules with a sharp point are found at the base of the petioles.

The 2-6-flowered umbel-like inflorescences are borne on short peduncles. The petals are pale purple or pink-purple with beetroot purple streaks on the upper two. These upper petals have an entire or slightly notched tip. Seven fertile stamens are present. Plants usually flower from August to December.

Distribution: The species is confined to the Southern and Eastern Cape where it occurs from near Barrydale eastwards to Engcobo in the Transkei. It is usually found among shrubs on mountain sides and often in ravines or kloofs near streams.

Remarks: *P. radens* is readily propagated by cuttings and was introduced into Britain in 1774. It is one of the pelargoniums being cultivated for oil of geranium.

This species is commonly known as *P. radula*, but Moore (1955) pointed out that this name was 'superfluous' when published and must therefore be rejected.

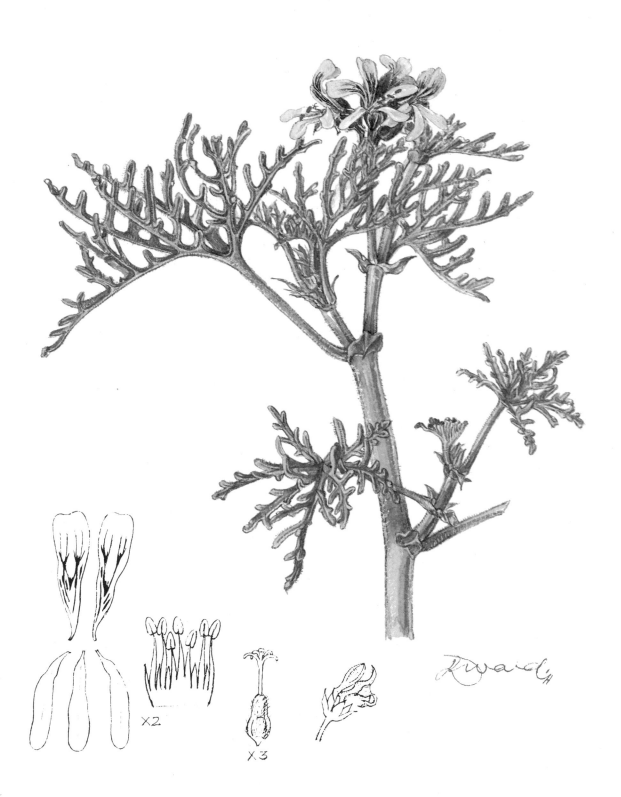

×2

×3

Pelargonium rapaceum

("Bergaartappel", "Bergpatat", "Norretjie").

Rapaceum (Latin), turnip-shaped; refers to the tuber resembling that of a turnip.

Acaulescent plant with a subterranean tuber.
Leaves linear and radical with numerous soft hairy leaflets.
Flowers pea-like with five fertile stamens.

P. rapaceum (L.) L'Herit. in Ait. Hort. Kew. ed. 1,2: 418 (1789); Knuth in Pflanzenr. 4,129: 348 (1912).
Originally described by Linnaeus as *Geranium rapaceum* in Syst. Nat.: 1141 (1759).

Description: An acaulescent (apparently stemless) plant with a subterranean tuber. It usually has a single large tuber although two or more heads can develop from it with age. The tuber can reach the size of a man's fist and is usually crowned with dead bases of old petioles. Besides this large tuber a string of spherical or elongated root tubers could also be formed.

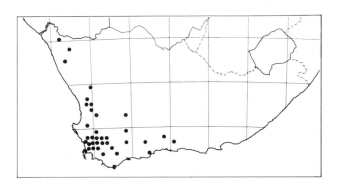

The erect and radical leaves are linear in outline and up to 40 cm long and 4,5 cm broad. They are much dissected (bi-pinnati-partite or bi-pinnatisect) with numerous soft hairy leaflets crowded in dense false whorls on the midrib. The leaflets are in turn divided into linear or broader segments. Awl-shaped stipules are found at the base of the petioles.

Long-peduncled and many-flowered umbel-like inflorescences appear from October to February at a stage when the leaves are dying. The flowers are most peculiar and fascinating. They resemble miniature pea-flowers in having the upper two petals bent back and the three lower ones held close together enclosing the anthers entirely. The petals are creamy, yellow, primrose-yellow or pink with red stripes on the lower part of the two upper petals. Five fertile stamens are present.

Distribution: This species occurs in Namaqualand, the Karoo, Eastern Cape but is very abundant in the South Western Cape. It grows in sandy soil and is often found on dry, stony mountain slopes.

Remarks: The Hottentots ate the tuber which they roasted in hot ashes. Reports are that the species has a medicinal use as an astringent antidiarrhoetic.
Masson introduced it into Britain in 1788.
Three varieties are distinguished on basis of the colour of the flowers.

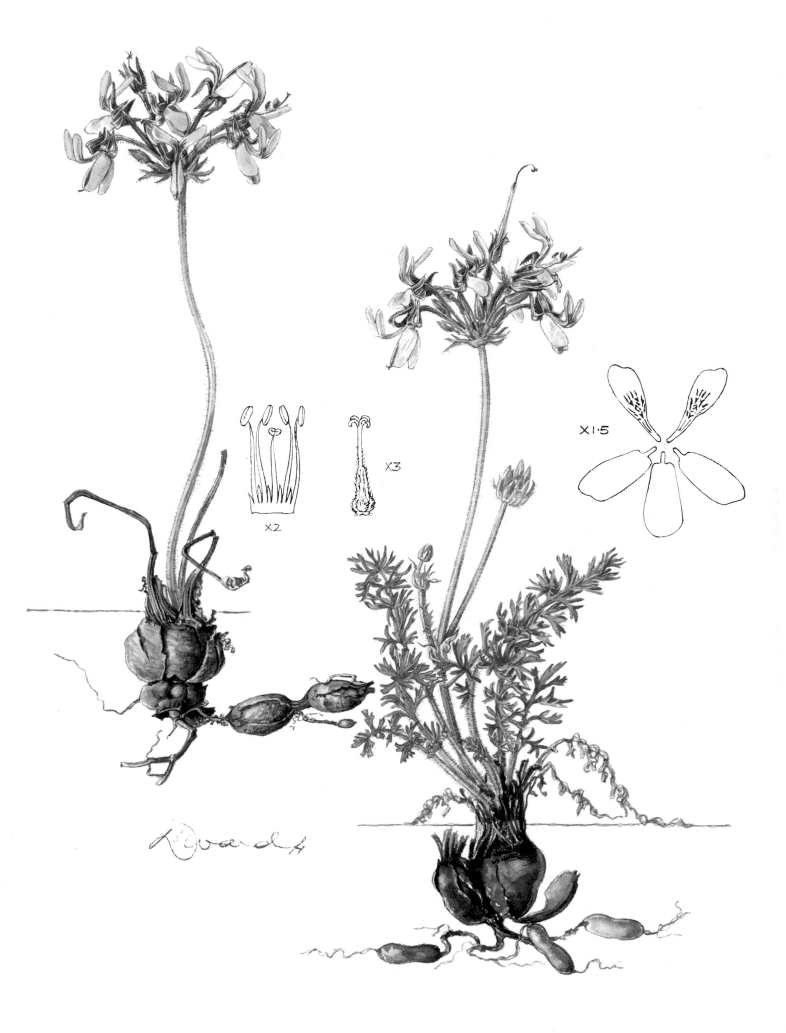

×2   ×3

×1·5

Pelargonium reniforme

(Kidney-leaved pelargonium, "Rooirabas")

Reniformis (Latin), Kidney-shaped; refers to the form of the leaf.

Small perennial shrublet with tuberous roots.
Leaves reniform or cordate, velvety and prominently veined underneath.
Flowers usually have a magenta colour.

P. reniforme Curt. in Bot. Mag. 14: t.493 (1800) — original description; Knuth in Pflanzenr. 4, 129: 447 (1912).

Description: A perennial erect shrublet with tuberous roots, which may attain a height of up to 1 m, but normally less than 40 cm tall. The relatively short main stem is often covered with the remains of old stipules. Flowering branches are relatively soft and slightly downy.
The reniform or cordate leaves with crenate or finely lobed margins are usually 2-3 cm in diameter. Leaves on the main stem are normally larger than those on flowering stems. Most leaves have a velvety texture and greyish-green colour, due to the presence of matted hairs. The lower side of the leaves shows a thicker hairiness and stronger venation than the upper side. Awl-shaped

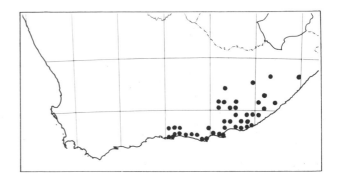

stipules are found at the base of the relatively long petioles.
Flowers are borne in 3- or more-flowered umbel-like inflorescences. The colour of the flowers varies from pink to magenta with a darker spot and stripes on the upper two petals. Six or seven fertile stamens are present. Plants flower throughout the year.

Distribution: This species occurs from Knysna eastwards to Umtata and is particularly common in the Eastern Cape. It grows on dry flats and in grassveld which is periodically subjected to burning.

Remarks: Harvey (1860) distinguished three varieties of this species on the basis of leaf characters and flower colour. The one variety with blackish-purple flowers is now considered as a separate species (*P. sidaefolium*).
The Xhosa people use the tuberous roots as a remedy for diarrhoea and dysentery. It is also used for liver complaints in sheep and calves. A piece of the root is tied along the bit in the mouth of a horse to prevent purging while on the road.
P. reniforme is a very attractive plant, especially when growing in groups. Cuttings root freely when planted in an equal mixture of very sandy loam and peat.

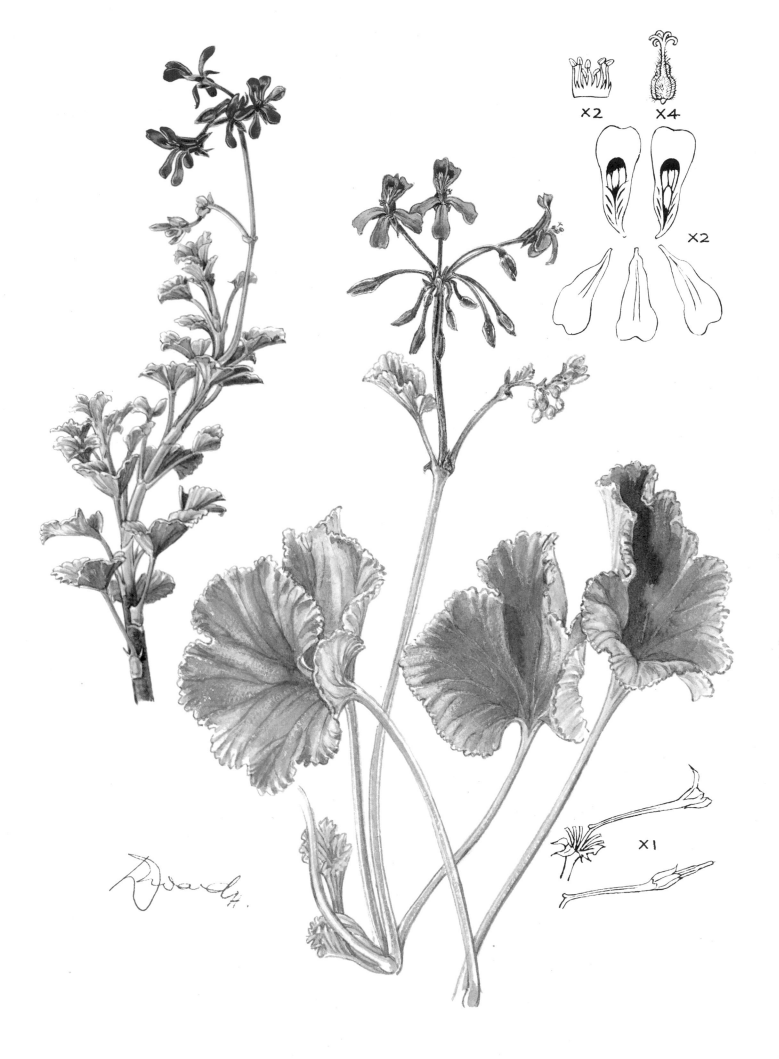

×2 ×4

×2

×1

Pelargonium rhodanthum

Rhodanthus (Latin), rosy-flowered

Perennial shrublet without spiny stipules.
Leaves usually cordate-reniform, relatively
small, grey-green.
Flowers attractive, pink to deep magenta
with dark spots.

P. rhodanthum Schltr. in Bot. Jahrb. 27: 152
(1900) original description; Knuth in
Pflanzenr. 4,129: 451 (1912).

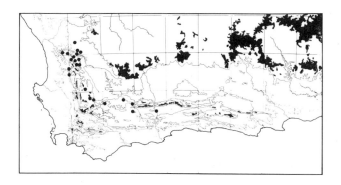

Description: A perennial shrublet with tuberous roots, attaining a height of 1 m although
normally smaller. The branches are rather
woody, almost glabrous and without persistent spine-like stipules.
The cordate-roundish or cordate-reniform
leaves with crenate or crenulate margins are
usually obscurely 3-lobed. Leaves on
branches of the main stem are 0,5-1 cm long
and 0,6-1,5 cm broad, but those on the flowering stem are normally much smaller. The
somewhat crisped leaf blades have a
greyish-green colour and they are covered
with shortish hairs (sub-tomentose). Ovate,
non-persistent stipules are found at the base
of the relatively long petioles.
Flowers are borne in 2-9-flowered umbel-like
inflorescences. The colour of the attractive
flowers varies from pink to mauve to deep
magenta with very dark spots and stripes on
all five petals. All the petals are shallowly
notched at their apex. Six or seven fertile
stamens are present. Plants flower from May
to October.

Distribution: This species is restricted to the
drier parts of the South Western Cape Province, where it occurs from Van Rhynsdorp
to Montagu and eastwards to the Calitzdorp
district. It usually grows on rocky outcrops.

Remarks: A new hybrid of *Pelargonium* was
described and named as *P. rhodanthum* by
Sweet in 1825. According to the International
Rules of Botanical Nomenclature, *P. rhodanthum* of Schlecter must be regarded as a
later homonym, and thus an illegitimate
name. Unfortunately *P. rhodanthum* Schltr.
must receive a completely new name since
no synonyms exist.

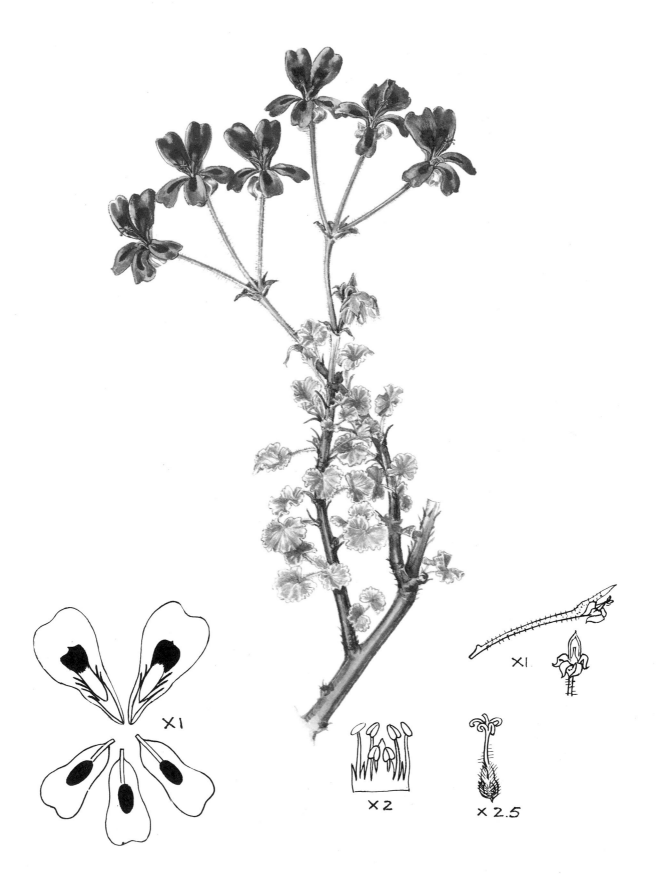

Pelargonium scabrum

(Rough-leaved pelargonium; Three-pointed pelargonium)

Scaber (Latin), scabrous; refers to the leaves and stems which are rough to the touch.

Erect and much branched shrub.
Leaves angularly 3-lobed, scabrous.
Flowers small, white, pink or mauve

P. scabrum (Burm. f.) L'Herit. in Ait. Hort. Kew. ed. 1,2: 430 (1789); Knuth in Pflanzenr. 4, 129: 462 (1912).
Originally described as *Geranium scabrum* by Burman (filius) in Spec. Bot. Ger.: 34 (1759).

Description: A rather woody, erect and much-branched shrub attaining a height of up to 2 m. The stems are scabrous due to the presence of rigid hairs.
Bruised leaves have a characteristic lemon scent, although this feature varies with locality. The leaves are rhomboidal in outline, angularly 3-lobed, palmately veined and normally ca. 4,3 cm long and ca. 4 cm broad. Plants growing further inland in the Western Cape have relatively more narrow-lobed leaves. The leaves may vary from

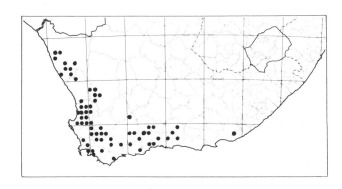

being almost glabrous to scabrous with glands always present. The leaf margin is finely to coarsely toothed. Deltoid stipules, often red-coloured, are found at the base of the harsh petioles. With age, the stipules wither to a papery texture.
Relatively small flowers are borne in 2-6 umbel-like inflorescences. The peduncles may be branched. The two upper petals are twice as large as the lower three. Their colour varies from white to dark pink or even mauve, with purple stripes on the upper two petals. Seven stamens bear anthers. Flowers are borne from May to January.

Distribution: It occurs from Springbok along the western coastal region, to the South Western Cape, and eastwards as far as Grahamstown. This species usually grows in sandy soil in dry habitats.

Remarks: A natural hybrid between *P. scabrum* and *P. glaucum* has been named *P. tricuspidatum.* It is difficult to propagate this pelargonium vegetatively as the cuttings do not strike readily.
Kennedy and Lee introduced it to England in 1775.

X1.

X2

X2

Pelargonium schizopetalum

(Divided-petalled or Orchid pelargonium; "Muishondbossie")

Schizo- (Greek), split, deeply divided; -petalum (Latin), petal.

Geophyte, stem very short and succulent. Leaves oblong ovate, pinnately incised or lobed, hairy.
Petals fimbriate, spur long.

P. schizopetalum Sweet in Geraniaceae 3: t.232 (1824) — original description; Knuth in Pflanzenr. 4, 129: 367 (1912).

Synonym: *P. amatymbicum* (Eckl. & Zeyh.) Harv. in Fl. Cap. 1: 277 (1860).

Description: A geophyte with a subterranean, oblong tuber which is often branched. The rather succulent stem is very short, while the conspicuous inflorescence may attain a height of 50 cm or more.
The dark green, radical leaves are glandular and variably covered with long hairs. The oblong ovate leaf blades with a cordate base are shallow to deeply pinnately incised or lobed, their margin being unevenly crenate. They are slightly fleshy, 5-16 cm long and 3-11 cm broad, and borne on relatively thick petioles which are flattened or chanelled on

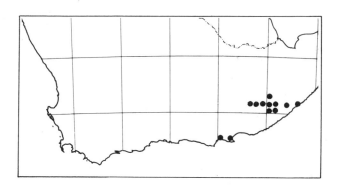

the upper side. Lanceolate stipules are fused to the base of the petioles.
The long unbranched scape, covered with coarse hairs, terminates in a 5-20-flowered pseudo-umbel. A fascinating feature of this species is its flower, which has fimbriate petals and a long, prominent spur. The colour of the five equally-sized petals varies from pale yellow to yellow-green to brown-purple, and they are striped with red or purple. Seven filaments bear anthers. It flowers from October to February.

Distribution: This interesting pelargonium is apparently restricted to a relatively small area in the Eastern Cape and Transkei, occurring from Port Elizabeth to Somerset-East and eastwards to Kentani.
It is commonly found on grassy slopes in damp locations.

Remarks: In the past *P. schizopetalum* and *P. amatymbicum* have been regarded as different species. *P. schizopetalum* is supposed to have less flowers per inflorescence, and more deeply incised leaves with a denser pubescence, than *P. amatymbicum* (leaves of the latter are illustrated). Intermediate forms occur and it seems likely that they are conspecific.
The flowers are unpleasantly scented — hence the Afrikaans vernacular name "Muishondbossie".
Colvill was the first to cultivate this curious plant in England in 1821.

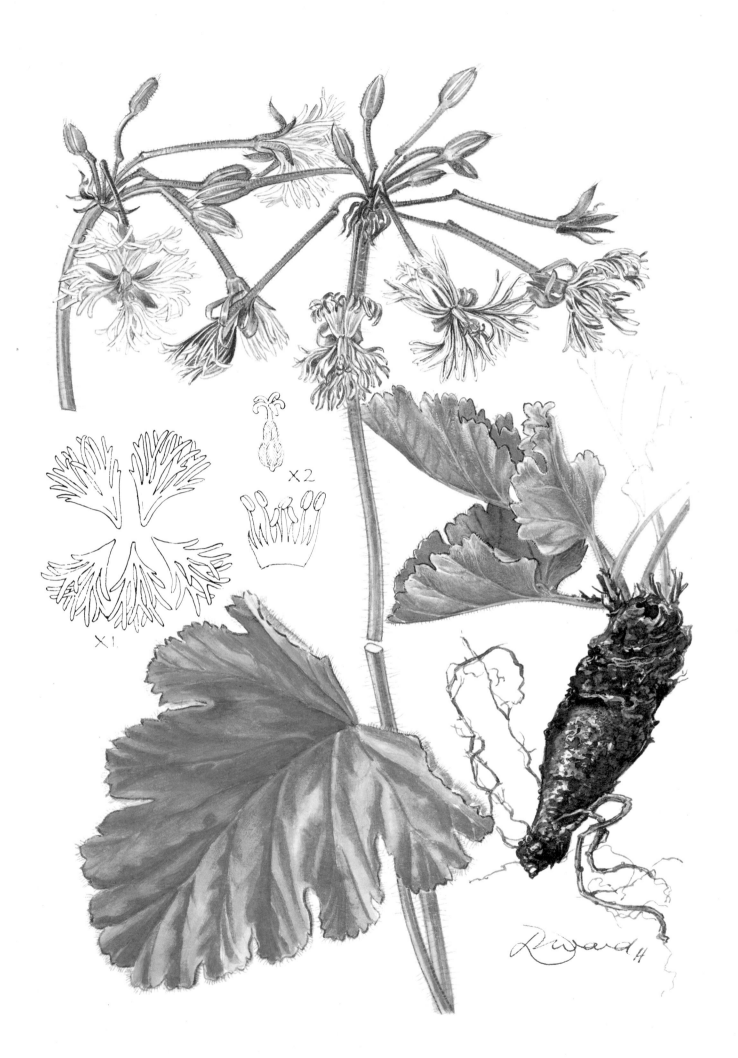

X2

X1

Pelargonium tabulare

(Table Mountain pelargonium)

Tabularis (Latin), flattened horizontally; it refers to Table Mountain where this species occurs.

Diffuse herb, stems and leaves glandular. Leaves kidney-shaped, stipules lanceolate. Flowers white to flesh-coloured.

P. tabulare (L.) L'Herit. in Ait. Hort. Kew. ed. 1,2: 419 (1789); Knuth in Pflanzenr. 4,129: 431 (1912).
Originally described by Linnaeus as *Geranium tabulare* in Sp. Pl. ed. 2,2: 947 (1763).

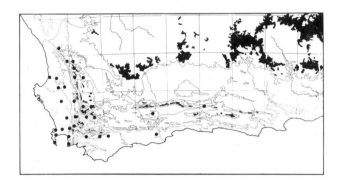

Description: A diffuse, often many-stemmed herb usually 20-30 cm high. The erect or straggling stems are herbaceous and covered with long glandular and rather coarse non-glandular hairs (hirsute). The side-branches often die during dormant periods giving the plants a stunted appearance.

The leaves are mostly kidney-shaped in outline with an angular-cordate base. They are 5-7-palmately lobed, usually 1,5-2,5 cm long and 2-4 cm broad. The surface of the leaf blades is sparsely hirsute but the veins and toothed margins of the rounded lobes are more densely haired. Leaves are often zoned, being marked with a purple to brown circle or horse-shoe. Exceptionally long petioles are characteristic of this species. They are usually 4-9 cm long and have a similar indumentum as the stems. Lanceolate stipules with hairy margins are present.

The umbel-like inflorescences with hirsute peduncles and 1-6 flowers each, resemble those of *P. alchemilloides.* Most flowers are white or cream-coloured, though flesh-coloured ones occur. Although similar in length, the two upper petals are usually broader than the three lower ones. Seven fertile stamens are present. It blooms in Spring but flowers can be found throughout the year.

Distribution: This species is restricted to the south western part of the Cape Province and occurs from Van Rhynsdorp to near Uniondale. It is often found on hill-slopes and disturbed areas.

Remarks: *P. tabulare* and *P. alchemilloides* are closely allied species and they have the same habit. *P. alchemilloides,* however, lacks long glandular hairs on the stems and leaves and the stipules of this species are ovate to broadly ovate.

×1.5

×3

×2

×1

Pelargonium tetragonum

(Square-stemmed pelargonium).

Tetra- (Greek), four, gonia- (Greek) angled; refers to the square stem.

Stems angular and fleshy.
Leaves small and fleshy.
Flowers large with four petals and bent filaments.

P. tetragonum (L.f.) L'Herit. in Ait. Hort. Kew. ed. 1,2: 427 (1789); Knuth in Pflanzenr. 4,129: 390 (1912).
Originally described as *Geranium tetragonum* by the son of Linnaeus in L.f., Suppl.: 305 (1781).

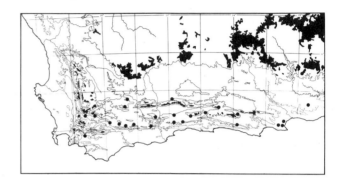

Description: A succulent xerophytic bush branching from the base and often sprawling in other plants attaining a height of up to 2 m. The thin and brittle stems are obtusely 4-3-angled, succulent, smooth and distinctly jointed.
Another xeromorphic character of this species is the relatively small and fleshy leaves. The leaves are normally ca. 2,5 cm long and ca. 4 cm broad, cordate or kidney-shaped, palmately-veined and crenately-lobed. The hairiness of the leaves varies from almost glabrous to villous, the margins often being fringed with long, soft hairs.

Relatively small and papery stipules are found at the base of the smooth or villous petioles.
The prominent flowers are very attractive, occuring in pairs on the peduncle. Only four petals are present, the upper two being much larger than the lower pair. The colour of the petals varies from cream to pale pink with deep red streaks on the upper two. Filaments of the seven fertile stamens are bent in the middle. It flowers from September to December.

Distribution: Occurring in a strip parallel with the coast from the Worcester-Caledon districts eastwards to Grahamstown. It has also been collected further inland near Graaff Reinet and Bedford.
This species is confined to dry habitats, usually growing with karroid vegetation on rocky outcrops.

Remarks: This is a remarkable and peculiar pelargonium with many extra-ordinary characters. Curtis (1791) summarized the uniqueness of the species as follows: "a vein of singularity runs through the whole of this plant, . . ."
Requirements for its cultivation are the same as for the more common pelargoniums being readily propagated by cuttings. Introduced to the Botanical Gardens at Kew by Masson in 1774.
Heavily grazed plants have been observed.

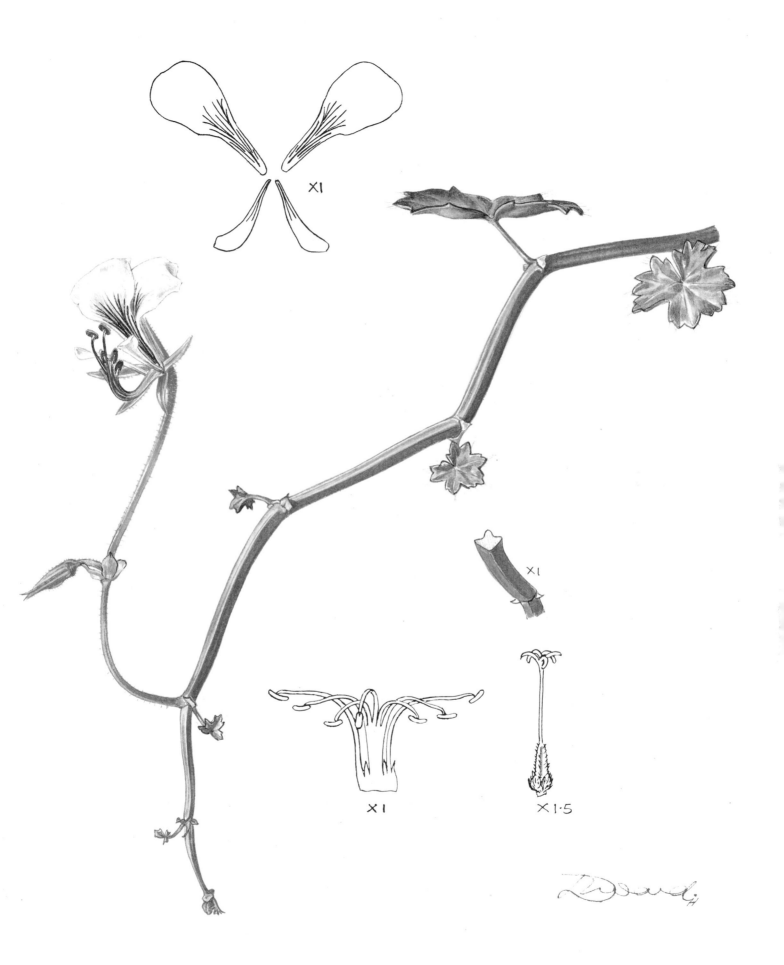

X1

X1

X1

X1.5

Pelargonium triste

(Night-scented pelargonium; "Kaneeltjie"; "Rooiwortel")

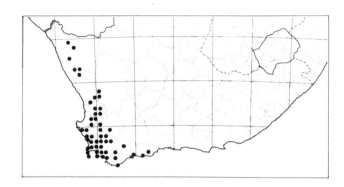

Tristis (Latin), dull coloured; refers to the dull colour of the flowers.

Geophyte with a large subterranean tuber.
Leaves like those of a carrot, hairy.
Flowers night-scented (musk), yellow to purplish.

P. triste (L.) L'Herit. in Ait. Hort. Kew. ed. 1,2: 418 (1789); Knuth in Pflanzenr. 4, 129: 358 (1912).
Originally described by Linnaeus as *Geranium triste* in Sp. Pl. ed. 1,2: 676 (1753).

Description: A geophyte with a large subterranean tuber and tuberous roots. The main stem is usually short and succulent. Before dormancy all the epiterranean parts of the plants die off.
The radical leaves resemble those of a carrot. They vary considerably in hairiness and structure (bi-tripinnately decompound or pinnatifid) with the segments decurrent, toothed and lacinate. The leaf blades are oblong ovate in outline and 10-45 cm long and 4-15 cm broad. Leaves on the flowering stems are usually much smaller than those on the main stem. Large cordate stipules are found at the base of the hairy petioles.
This species has the typical star-like and night-scented inflorescence of the section Polyactium. Pseudo-umbels with 6-20 flowers, are borne on very long and hairy peduncles. The petals, of a dull coloured yellowish-green to brownish-purple hue, are edged with a lighter margin. Seven fertile stamens are present. Plants flower from August to February.

Distribution: *P. triste* occurs in a strip running parallel to the coast from Steinkopf in the North-Western Cape to Albertinia in the Southern Cape. It usually grows in sandy soil. A subterranean network often gives rise to a population of separate plants.

Remarks: Three varieties of *P. triste* have been distinguished on the basis of leaf characters. It has been observed, however, that leaf structure and flower colour vary considerably in the same population. *P. flavum* which is considered a separate species, is probably conspecific with *P. triste.*
When John Tradescant took *P. triste* to England in 1632, it became one of the first pelargoniums to be transplanted from the Cape. A decoction from the scarlet tuber is used as an astringent to arrest diarrhoea or dysentery.

×2

×3

×1

×2

Pelargonium urbanum
var. bipinnatifidum

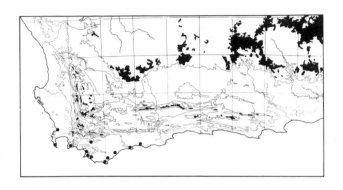

Urbanus (Latin), pertaining to towns and cities. The species was apparently first collected in an urban area. Bi- (Latin), two; pinnatifidus (Latin), pinnately cleft.

Sprawling halfshrubs, stems tomentose.
Leaves bipinnatifid, stipules large.
Flowers large, usually cream to yellow,
petals four or five.

P. urbanum (Eckl. & Zeyh.) Harv. var. *bipinnatifidum* Harv. in Fl. Cap. 1: 288 (1860); Knuth in Pflanzenr. 4, 129: 400 (1912). The species was originally described by Ecklon and Zeyher in Enum. 1: 71 (1835) as *Myrrhidium urbanum.*

Description: A perennial and sprawling halfshrub, attaining a height of up to 0,3 m, with tuberous roots. The stems are covered with dense woolly hairs (tomentose).
The leaves are ca. 4 cm long and ca. 2,5 cm broad, elongate cordate and bipinnatifid. The margin of the segments is dentate, and the leaves are variably hirsute to villous. Large, broadly ovate stipules are found at

the base of ca. 3,5 cm long tomentose petioles.
The peduncle bears 3-6 flowers in an umbel-like inflorescence. The flowers are large and very attractive. Four or five petals are present, the upper two being much larger than the lower three (or two). The colour of the petals varies from cream to pale yellow or pale pink with red streaks on the upper two. Seven fertile stamens are present. This variety flowers from October to January.

Distribution: It occurs along the coastal areas from Melkbosch Strand in the west to the coastal limestone ridges south of Riversdale.
This variety is confined to sandy areas often growing amongst clumps of *Thamnochortus* spp. and *Willdenowia striata* in coastal scrub.

Remarks: In 1860 Harvey distinguished two varieties within this species. They are var. *pinnatifidum* and var. *bipinnatifidum*. Var. *pinnatifidum* occurs mainly in the Eastern Province, it has pinnatifid leaves and dark purple-pink flowers.

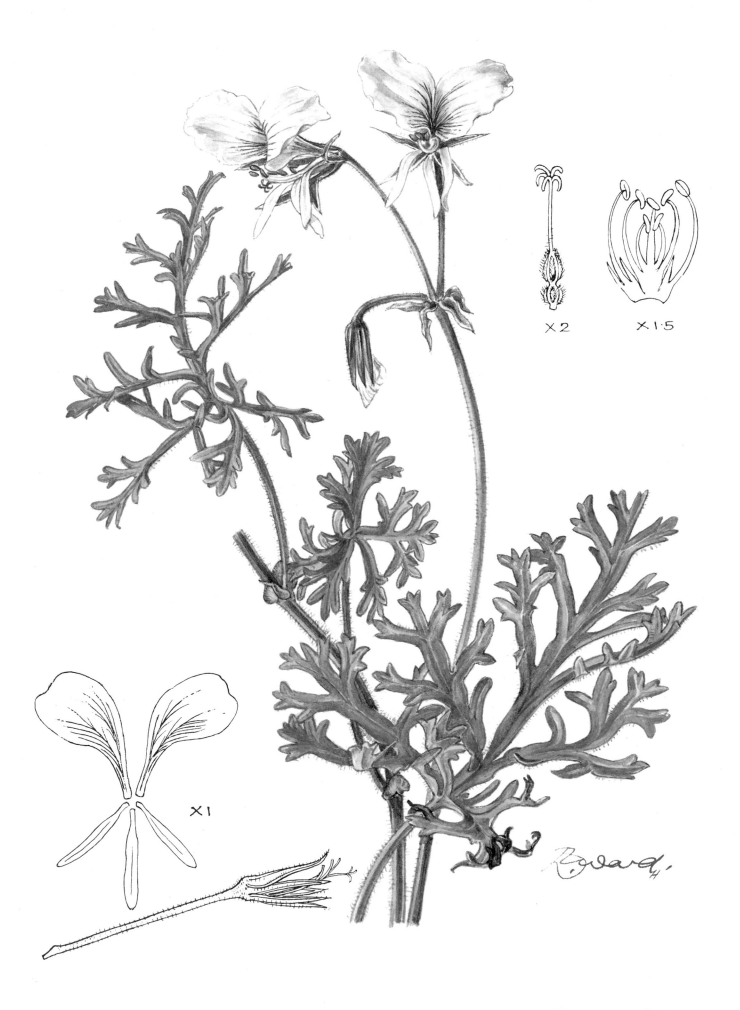

×2 ×1·5

×1

R.Ward.

Pelargonium violareum

Violareum — associated with *Viola tricolor*, the European wild pansy.

Halfshrub, main stem short and diffusely branched.
Leaves usually lanceolate, villous, tending to grey.
Flowers three-coloured, 3-5 fertile stamens.

P. violareum Jacq. in Icon. Pl. Rar. 3,10: t. 527 (1792) — original description.

Synonym: *P. tricolor* Curt. in Bot. Mag. 7: t.240 (1793); Knuth in Pflanzenr. 4, 129: 421 (1912).

Description: A perennial halfshrub with a maximum height of 30 cm. Plants tend to have an alpine habit due to the short main stem and diffuse side branches; these are covered with the remains of old stipules.

The leaves are narrowly lanceolate to ovate or obovate with an unevenly incised or toothed margin. The teeth are typically red-tipped. Some leaves have two distinctly larger lobes at their base. The length of the leaf blades varies from 1-4,5 cm and the width from 0,2-1,4 cm. They are covered on both sides with appressed, white hairs

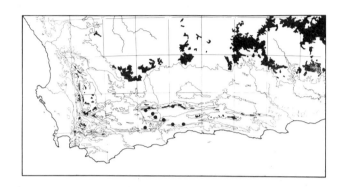

directed towards the base of the leaf. Their grey-green (canescent) colour is due to the variable hairiness of the leaf surface. Stipules are subulate, ca. 4 mm long, persistent and red-brown in colour.
The plants are usually very floriferous, as up to four peduncles, each bearing 2-4 flowers, sprout from the apex of most of the branches. The colour of the two upper petals shades from white to dark red. Each has a raised, warty, shiny dark spot at the base. The three lower petals are white, often with a narrow red streak extending from the base. Three to five filaments bear relatively small and dark-coloured anthers. The species flowers from September to December.

Distribution: *P. violareum* occurs on the Swartberge near Ladismith and Oudtshoorn, on Rooiberg near Van Wyksdorp and on the Langeberg near Riversdale.
It grows in sandy soil in False Macchia.

Remarks: Curtis (1793) stated that on watering these plants, the substance in the warty, dark spot on the upper petals dissolves, staining the petals a red colour. This phenomenon has also been observed in plants in the wild.
This species was cultivated in England as early as 1792 and has always been appreciated for the beauty of its flowers.

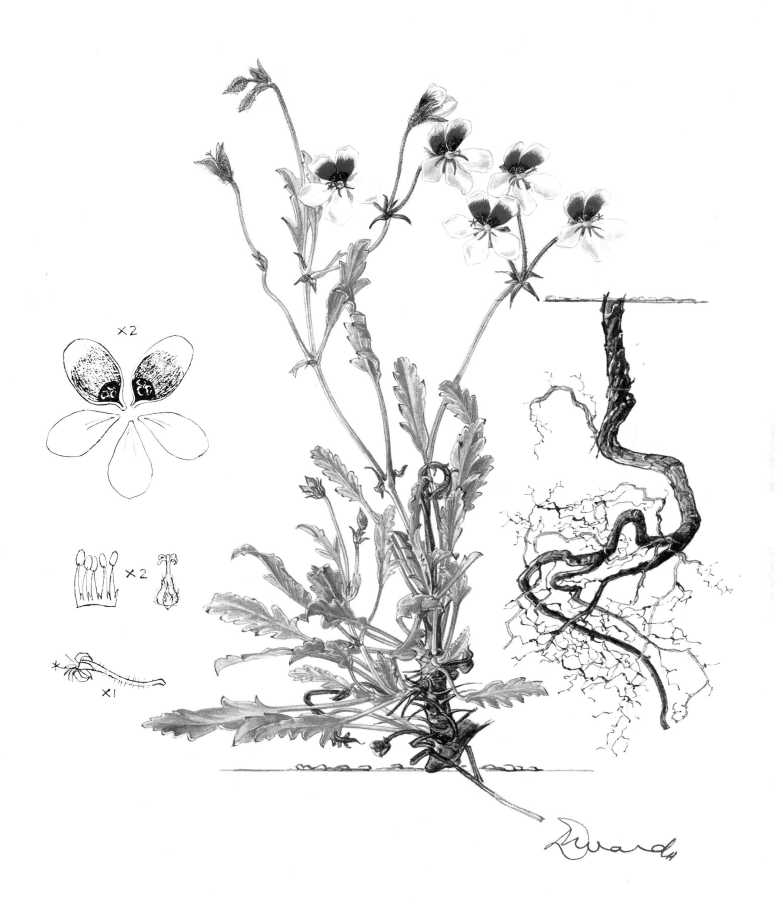

Pelargonium vitifolium

(Vine-leaved pelargonium; Balm-scented pelargonium)

Viti- (Latin), pertaining to the vine; -folium (Latin), leaf; refers to the leaves of the vine which it resembles.

Erect and strongly aromatic shrub.
Leaves cordate, 3-5-lobed, slightly rough to the touch.
Flowers pink to carmine, small capitate-like inflorescences.

P. vitifolium (L.) L'Herit. in Ait. Hort. Kew. ed. 1,2: 425 (1789); Knuth in Pflanzenr. 4,129: 468 (1912).
Originally described by Linnaeus as *Geranium vitifolium* in Sp. Pl. ed. 1,2: 678 (1753).

Description: An erect and strongly aromatic shrub, 0,5-1 m tall. The main stem is woody at its base, while the side branches are herbaceous and covered with long soft hairs (villous).
The cordate leaves are usually ca. 6 cm long and ca. 8 cm broad, mostly 3-lobed but occasionally 5-lobed and slightly rough to the touch, due to rather stiff hairs covering the leaf blades. They are more glandular and strongly scented than those of *P. capitatum,* but not crinkled. The margins of the leaves

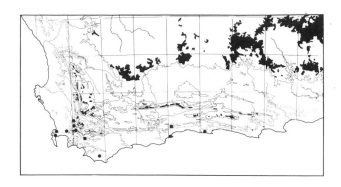

are coarsely and irregularly toothed. Broadly ovate stipules, sometimes 2- to 3-toothed, are found at the base of the petioles, which are usually longer than the leaf blades.
The capitate-like inflorescences resemble those of *P. capitatum* but they are less hairy, smaller and usually not more than 12-flowered. The colour of the petals varies from light pink to carmine with violet-purple stripes on the two larger upper petals. Seven fertile stamens are present. It is in full bloom during Spring (August-October), and except for the mid-winter months, flowers are found throughout the year.

Distribution: This species is restricted to the South Western and Southern Cape; occurring in the Peninsula and eastwards to Knysna. It grows amongst other shrubs in half-shaded localities near streams or rivulets.

Remarks: *P. vitifolium* is a vigorous grower and can easily be cultivated from cuttings. It was introduced to England in 1724 where it was grown in the famous Chelsea Garden. This species is closely related to *P. capitatum* and *P. papilionaceum.* They have many characters in common but are definitely not conspecific. The differences between the three species can be deduced from the diagnostic features given for each.

X2

X2

Pelargonium zonale

(Horse-shoe pelargonium, "Wilde malva").

Zona (Latin), zone; refers to the horse-shoe zone on the leaves.

An erect or scrambling softly woody shrub. Leaves almost orbicular, cordate, often smooth and zoned.
Flowers irregular, petals equally sized, staminal column relatively short.

P. zonale (L.) L'Herit. in Ait. Hort. Kew. ed. 1,2: 424 (1789); Knuth in Pflanzenr. 4, 129: 443 (1912).
Originally described by Linnaeus as *Geranium zonale* in Sp. Pl. ed. 1,2: 678 (1753).

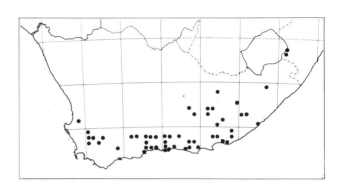

Description: An erect or scrambling and softly woody shrub usually ca. 1 m tall but sometimes as high as 3 m. The old stems harden with age but the young branches are almost succulent and usually covered with short hairs.
The almost orbicular leaves, with a cordate base, are usually 5-8 cm in diameter. They are obsoletely many-lobed with irregularly scalloped (crenate) margins. The leaves are glabrous or slightly pubescent with short hairs and glands. Zonation of the leaves by means of a dark horse-shoe mark on the upper surface is often found but this is not an invariable character. Broadly ovate stipules, present at the base of the long petioles soon become membranous and withered.
The inflorescences are typically umbel-like and 5-70 flowered with the flower buds reflexed. The flowers have a distinctly zygomorphic (irregular) shape due to the characteristic grouping of the petals into an upper pair and three lower ones. The equally sized petals are commonly rose-coloured with reddish stripes, but all shades of red to pure white petals may be found. Seven fertile stamens are present and the staminal column is relatively short in comparison with that of the closely allied *P. inquinans*. Flowers are found throughout the year.

Distribution: This well known species is widely distributed, occurring from Piketberg eastwards to the Underberg district in Natal and is particularly common in the coastal areas of the Southern Cape. It grows on the margin of indigenous forests or on rocky hill-sides with scrub vegetation.

Remarks: *P. zonale* is a parent of many widely cultivated hybrids. *P.* x *hortorum*, the common garden or zonal pelargonium, has been largely derived from hybridization between *P. zonale* and *P. inquinans*.
P. zonale was first cultivated in England by the Duchess of Beaufort in 1710.

X.1.

X.2

Index to the vernacular names

Glossary

Acaulescent: stemless, or apparently so.

Actinomorphic flower: with a regular or star pattern, capable of bisection in two or more planes into similar halves.

Acuminate: an acute apex with sides slightly concave and tapering to a protracted point.

Annual: of one season's duration from seed to maturity and death.

Anther: pollen-bearing part of the stamen.

Canescent: tending to grey, becoming grey or hoary.

Capitate: with a knob-like head.

Calyx: the outer whorl of floral envelopes, composed of the sepals.

Ciliate: fringed with hairs; bearing hairs on the margin.

Conspecific: belonging to the same species.

Cordate: heart-shaped, applied to leaves having the petiole at the broader and notched end.

Crenate: shallowly round-toothed or obtusely toothed, scalloped.

Crenulate: crenate but the toothings themselves small.

Cuneate: wedged-shaped; triangular with the narrow end at point of attachment.

Decumbent: reclining or lying on the ground, but with end ascending.

Decurrent: extending down and adnate to the stem.

Deltoid: triangular.

Dorsifixed: attached to the back of the anthers.

Endosperm: the nutritive tissue of seeds.

Entire: with a continuous margin; not in any way indented.

Epithet: the second part of a binomial (name of a species).

Filament: the stalk of an anther.

Filiform: thread-shaped; long and very slender.

Fimbriate: fringed.

Flexuose: bent alternately in opposite directions, zigzag.

Floriferous: flower-bearing.

Fynbos: bushy, macchia-like vegetation of the South Western Cape; plants characterized by fine and hard leaves.

Geophyte: plants with perennating organs (e.g. tubers) underground.

Glabrous: without any hairs.

Glaucous: sea-green; covered with a bloom as of a cabbage leaf.

Habitat: the kind of locality in which a plant grows.

Half-shrub: perennial plant with only lower part woody.

Herb: plants lacking a definite woody firm structure.

Herbaceous: referring to plants having the characteristics of herbs.

Hirsute: with rather rough or coarse hairs.

Homonym: any of two or more identical names based on different types, only one of which can be legitimate.

Hypanthium: an enlargement or development derived from the fusion of floral envelopes.

Imbricate: overlapping, as shingles on a roof.

Indumentum: any covering, e.g. hairiness.

Inflorescence: manner of flower bearing.

Internode: the part of an axis between two nodes.

Lacerate: torn; irregularly cleft or cut.

Laciniate: slashed into narrow pointed lobes.

Lanceolate: Lance-shaped; widening above the base and tapering to the apex; much longer than broad.

Lobulate: having small lobes.

Macchia: see Fynbos.

Mericarp: a portion of a fruit which splits off as a perfect fruit.

Obovate: the reverse of ovate, the terminal half broader than the basal half.

Ovate: shaped like a longitudinal section of a hen's egg, the broader end basal.

Ovoid: a solid that is oval in flat outline.

Palmate: lobed or divided or ribbed in a palm-like or hand-like fashion.

Panicle: an inflorescence, the main axis of which is branched and the flowers stalked.

Panicled: furnished with a panicle.

Papilionaceous: butterfly-like, pea-like flower.

Pedicel: stalk of a single flower.

Peduncle: stalk of a flower cluster.

Peltate: attached to its stalk inside the margin; peltate leaves are usually shield-shaped.

Perennial: a plant with a lifespan of several years.

Petiole: leaf stalk.

Pinnate: feather-formed; with the leaflets of a compound leaf placed on either side of the rachis.

Pinnately: in a pinnate fashion.

Pinnatifid: margin pinnately cut halfway to the midrib.

Pinnatipartite: margin pinnately cut or cleft not quite to the midrib.

Pinnatisect: margin pinnately cut down to the midrib.

Polymorphous: with several or various forms.

Procumbent: lying along the ground.

Prostrate: a general term for lying flat on the ground.

Pseudo-umbel: false umbel; the oldest flowers occupying the centre of the umbel.

Puberulous: minutely pubescent, the hairs soft, straight, erect, scarcely visible to the unaided eye.

Pubescence: the hairiness of a plant organ.

Pubescent: clothed with soft hair or down.

Rachis: axis bearing flowers or leaflets.

Radical: arising from the root or its crown.

Receptacle: torus, the more or less enlarged or elongated end of the stem or flower axis on which some or all of the floral parts are borne.

Reflexed: abruptly recurved or bent downward or backward.

Reniform: kidney-shaped.

Scabrous: rough; feeling roughish or gritty to the touch.

Schizocarp: a dry dehiscent fruit which splits into one-seeded portions, mericarps or "split-fruits".

Serrate: a margin which is saw-toothed with the teeth pointing forward.

Serrulate: minutely serrate.

Sessile: not stalked; sitting.

Specific: relating to a species.

Spur: a hollow and slender extension of some part of the flower, usually nectariferous.

Staminal column: a column formed by the fusion of the bases of filaments.

Staminode: a sterile stamen.

Stipule: a basal appendage of a petiole.

Stolon: a stem that grows horizontally along the ground surface.

Subulate: awl-shaped, tapering from base to apex.

Succulent: fleshy; soft and thickened in texture.

Synonym: a rejected name due to a misapplication or difference in taxonomic judgement.

Tomentose: densely woolly or pubescent; with matted soft wool-like hairiness.

Trifoliolate: a leaf with three leaflets.

Tuber: an enlarged plant organ, usually subterranean.

Undulate: wavy.

Villous: provided with long and soft, not matted, hairs; shaggy.

Xerophyte: a plant which can subsist with a small amount of moisture.

Zygomorphic flower: divisible into equal halves in one plane only.

Bibliography

Adamson, R.S. & T.M. Salter 1950. Flora of the Cape Peninsula. Cape Town: Juta & Co. Ltd.

Aiton, W. 1789. Hortus Kewensis Ed. 1, vol. 2. London.

Andrews, H.C. 1797-1815. The botanist's repository. London.

Andrews, H.C. 1805-1806. Geraniums vol. 1 & 2. London.

Boerhaave, H. 1720. Index alter plantarum quae in horto academico Lugduno-Batavo aluntur. Leiden.

Burman, J. 1738. Rariorum africanarum plantarum. Amsterdam.

Burman, N.L.B. (filius) 1759. Specimen botanicum de Geraniis. Leiden.

Burman, N.L.B. (filius) 1768. Flora Indica: cui accedit series zoophytorum indicorum nec non prodromus florae capensis. Amsterdam/Leiden.

Carolin, R.C. 1961. The genus Pelargonium L'Her. ex Ait. in Australia. *Proc. Lin. Soc. N.S.W.* 86, 3: 280-294.

Cavanilles, A.J. 1787. Monadelphiae classis dissertationes decem; Quarta dissertatio botanica, de Geranio. Paris.

Clifford, D. 1970. Pelargoniums including the popular 'Geranium'. London: Blanford Press.

Commelin, C. 1706. Horti medici Amstelaedamensis plantae rariores et exoticae. Leiden.

Curtis, W. 1791-1800. The botanical magazine. London.

De Candolle, A.P. 1824. Prodromus systematis naturalis regni vegetabilis, vol. 1. Paris.

Dietrich, F.G. 1807. Vollständiges Lexicon der Gärtnerei und Botanik, vol. 7. Berlin.

Dillenius, J.J. 1732. Hortus elthamensis. London.

Don, G. 1831. A general system of gardening and botany, vol. 1. London.

Dyer, R.A. 1953. The flowering plants of Africa, vol. 29. Pretoria: The Government Printer.

Dyer, R.A. 1974. The genera of Southern African flowering plants. Pretoria: The Government Printer.

Ecklon, C.F. & K.L. Zeyher 1835. Enumeratio plantarum africae australis extratropicae, vol. 1. Hamburg.

Harvey, W.H. 1860. Flora capensis, vol. 1. Dublin: Hedges, Smith and Co.

Hermann, P. 1687. Horti academici Lugduno-batavi catalogus. Leiden.

Hermann, P. 1689. Paradisi batavi prodromus. Amsterdam.

Hutchinson, J. 1969. Evolution and phylogeny of flowering plants. London: Academic Press.

Jacquin, N.J. 1781-1795. Icones plantarum rariorum, vol. 1-3. Vienna.

Jacquin, N.J. 1786-1796. Collectanea, vol. 1-5. Vienna.

Jacquin, N.J. 1797-1804. Plantarum rariorum horti caesarei schoenbrunnensis, vol. 1-4. Vienna.

Johnson, T. 1633. The Herbal of Gerard. London.

Karsten, M.C. 1951. The old Company's garden at the Cape and its superintendents. Cape Town: Maskew Miller Limited.

Knuth, R. 1912. Das Pflanzenreich 4, 129. Berlin.

Kokwaro, J.O. 1969. Notes on East African Geraniaceae. *Kew Bull.* 23: 527-530.

Leighton, F.M. 1932. *Pelargonium hollandii. S. Afr. Gdng. Country Life* 22: 232.

L'Heritier de Brutelle, C-L. 1789. Compendium Generalogium exhibens Erodium, Pelargonium, Geranium, Monsoniam et Grielum. Unpublished manuscript.

L'Heritier de Brutelle, C-L. 1792. Geraniologia, sue Erodii, Pelargonii, Monsoniae et Grieli historia iconibes illustrata. Paris.

Linnaeus, C. 1753. Species plantarum. Stockholm.

Linnaeus, C. 1755. Centuria plantarum I. Uppsala.

Linnaeus, C. 1759. Systema naturae. Stockholm.

Linnaeus, C. (filius) 1781. Supplementum plantarum. Braunschweig.

Mastalerz, J.W. 1971. Geraniums. A manual on the culture, diseases, insects, economics, taxonomy and breeding of geraniums. Pennsylvania: Pennsylvania Flower Growers.

Merxmüller, H. & A. Schreiber 1966. Prodromus einer Flora von Südwest Afrika 4, 64. Lehre: Verlag von J. Cramer.

Moore, H.E. 1955. Pelargoniums in cultivation. *Baileya* 3,1: 5-47.

Müller, T. 1963. Flora Zambesiaca 2,1. London: Crown Agents for oversea governments and administrations.

Schlechter, R. 1900. Plantae schlechterianae novae vel minus cognitae describuntur. *Bot. Jahrb.* 27: 86-220.

Sims, J. 1802. The botanical magazine, vol. 15. London.

Smith, J.E. 1793. Icones pictae plantarum rariorum. London.

Sprengel, K.P.J. 1826. Caroli Linnaei Systema vegetabilum, vol. 3. Göttingen.

Steudel, E.G. 1841. Nomenclator botanicus, vol. 2. Stuttgart.

Sweet, R. 1820-1830. Geraniaceae, vol. 1-5. London.

Sweet, R. 1822. A catalogue of plants sold by Colvill & Son. London.

Thunberg, C.P. 1794. Prodromus plantarum capensium. Uppsala.

Van. Royen, A. 1740. Florae Leydensis. Prodromus exhibens plantas quae in horto academico Lugduno-Batavo aluntur. Leiden.

Vorster, P. 1973. The flowering plants of Africa, vol. 42. Pretoria: The Government Printer.

Webb, W.J. 1971. Pelargonium species. Yearbook of the British Pelargonium and Geranium society.

Webb, W.J. 1972. More about Pelargonium species. Yearbook of the British Pelargonium and Geranium society.

Willdenow, C.L. 1800. Caroli a Linné Species plantarum, Editio quarta, vol. 3. Berlin.

Willdenow, C.L. 1804. Hortus berolinensis, vol. 1,2. Berlin.